Schriftenreihe der Professur für Molekulare Lebensmitteltechnologie

Band 6

Characterization of selected Polyphenol-Protein Interaction Products in alkaline-treated Sunflower Meal

Dissertation

zur Erlangung des Grades

Doktor der Ernährungswissenschaften (Dr. troph.)

der Landwirtschaftlichen Fakultät

der Rheinischen Friedrich-Wilhelms-Universität Bonn

von

Verena Bongartz

aus Konstanz

Bonn 2019

Angefertigt mit Genehmigung der Landwirtschaftlichen Fakultät der Universität Bonn

Referent: Prof. Dr. Andreas Schieber

Institut für Ernährungs- und Lebensmittelwissenschaften
Universität Bonn

Korreferent: Prof. Dr. Karl-Heinz Südekum

Institut für Tierwissenschaften
Universität Bonn

Tag der mündlichen Prüfung: 31. Mai 2019

Erscheinungsjahr: 2020

Bibliografische Information der Deutschen Nationalbibliothek

Die Deutsche Nationalbibliothek verzeichnet diese Publikation in der Deutschen Nationalbibliografie; detaillierte bibliographische Daten sind im Internet über http://dnb.d-nb.de abrufbar.

1. Aufl. - Göttingen: Cuvillier, 2020

Zugl.: Bonn, Univ., Diss., 2019

© CUVILLIER VERLAG, Göttingen 2020

Nonnenstieg 8, 37075 Göttingen

Telefon: 0551-54724-0

Telefax: 0551-54724-21

www.cuvillier.de

Alle Rechte vorbehalten. Ohne ausdrückliche Genehmigung des Verlages ist es nicht gestattet, das Buch oder Teile daraus auf fotomechanischem Weg (Fotokopie, Mikrokopie) zu vervielfältigen.

1. Auflage, 2020

Gedruckt auf umweltfreundlichem, säurefreiem Papier aus nachhaltiger Forstwirtschaft.

ISBN 978-3-7369-7180-6

eISBN 978-3-7369-6180-7

List of publications and conferences

Publications

Bongartz, V.; Brandt, L.; Gehrmann, M.L.; Zimmermann, B.F.; Schulze-Kaysers, N.; Schieber, A (2016): Evidence for the formation of benzacridine derivatives in alkaline-treated sunflower meal and model solutions. Molecules 21, 91.

Bongartz, V.; Böttger, C.; Wilhelmy, N.; Schulze-Kaysers, N.; Südekum, K.-H.; Schieber, A. (2018): Protection of protein from ruminal degradation by alkali-induced oxidation of chlorogenic acid in sunflower meal. J Anim Physiol Anim Nutr 102, 209.

Conferences

Bongartz, V.; Brandt, L.; Gehrmann, M.L.; Zimmermann, B.F.; Schulze-Kaysers, N.; Schieber, A (2016): Evidence for the formation of benzacridine derivatives in alkaline-treated sunflower meal and model solutions. Food-Nutrition-Health Symposium in Kiel, Germany, 12–13 May, 2015 [Poster presentation]

Declaration of contribution as co-author

The following co-authors contributed to the papers presented in Chapters 2 and 3:

Prof. Dr. Andreas Schieber contributed to the publications as the supervisor of this thesis and by proofreading all manuscripts.

Prof. Dr. Karl-Heinz Südekum assisted in the interpretation and publication of the results. As the corresponding author of chapter 3 he proofread the manuscript and handled the formal aspects of this publication.

Dr. Nadine Schulze-Kaysers assisted in the interpretation and publication of the results and by proofreading all manuscripts. As the corresponding author of chapter 2 she handled the formal aspects of this publication.

Dr. Benno F. Zimmermann assisted in the interpretation and publication of the results.

Christian Böttger assisted in the design of experimental work as well as the interpretion and publication of the results.

Nora Wilhelmy, **Lisa Brandt** and **Mai Linh Gehrmann** assisted in the design of experimental work.

Table of Contents

Chapter 1: General Introduction 1
   1   The sunflower (*Helianthus annuus* L) 1
      1.1   Origin and proliferation 1
      1.2   Botanic characteristics 2
      1.3   Oil production 2
   2   Solvent-extracted sunflower meal 5
      2.1   Composition 6
      2.2   Protein 6
      2.3   Phenolic content 7
   3   Polyphenols 9
      3.1   General overview 9
      3.2   Chlorogenic acid in SEM 11
   4   Polyphenol-protein interactions 13
      4.1   Non-covalent interactions 16
      4.2   Covalent interactions 17
   5   Trihydroxy benzacridine derivates 21
   6   Aim 22

Chapter 2: Protection of protein from ruminal degradation by alkali-induced oxidation of chlorogenic acid in sunflower meal 31
   1   Introduction 33
   2   Materials and methods 35
      2.1   Alkali treatment of sunflower meal 35
      2.2   Chemical analysis 35
      2.3   Crude protein fractionation and estimation of ruminally undegraded crude protein 36
      2.4   Intestinal digestibility of ruminally undegraded crude protein .. 37
      2.5   Data treatment and statistical analysis 38
   3   Results 39
      3.1   Chemical composition 39
      3.2   Color development 39
      3.3   Statistical relationships 39
      3.4   Crude protein fractions 40
      3.5   Ruminally undegraded crude protein 41

	3.6	Optimization and verification of predictive models	42
	3.7	Estimation of intestinal digestibility	43
4		Discussion	45
5		Conclusion	48

Chapter 3: Evidence for the Formation of Benzacridine Derivatives in Alkaline-Treated Sunflower Meal and Model Solutions 53

1 Introduction 54
2 Results and Discussion 56
    2.1 Color Development in SEM Extract and Model Solutions 56
    2.2 UHPLC-DAD-MS/MS Analysis 58
3 Experimental Section 65
    3.1 Plant Material 65
    3.2 Chemicals and Reagents 65
    3.3 UHPLC-DAD-MS/MS 66
4 Conclusions 68

Chapter 4: Concluding remarks 71

List of abbreviations 81

Summary 83

Zusammenfassung 85

Chapter 1: General Introduction

1 The sunflower (*Helianthus annuus* L)

1.1 Origin and proliferation

Helianthus (formerly Annui) is a genus of plants that contains approximately 70 species of sunflower in the Asteraceae. Although most species are perennial, section *Helianthus* includes more than 10 species most of which are annuals. The domesticated sunflower is derived from the wild form of *Helianthus annuus* L, or common sunflower [Schilling et al. 1998; Roth & Kormann, 2000].

Helianthus annuus L is native to North America. The wild form occurs throughout the continental United States, southern Canada, and northern Mexico. However, its prehistoric distribution is poorly understood. It is speculated that, prior to the arrival of Homo sapiens, the species was restricted to the south western United States. Native Americans used wild forms of *Helianthus annuus* L as a food, suggesting it became a camp-following weed and was thereby introduced to the central and eastern United States, where it was then domesticated [Heiser, 1951; Heiser et al., 1969].

Another theory states that buffaloes were the primary dispersal agents and that wild *Helianthus annuus* L had been widely distributed before the colonization of North America by humans. All modern domesticated sunflowers can be traced to a single center of domestication in eastern North America predating 3000 B.C. [Asch, 1993; Smith, 2011].

In the late 16th century, *Helianthus annuus* L – alongside beans and corn – was introduced to Europe by Spanish explorers, where it was cultivated as an ornamental plant. The commercial cultivation as an oil crop first started in Russia, around 1830. It resulted in high-oil lines with increasing oil contents from 28% to almost 50%. Today, Helianthus annuus L is cultivated throughout North and South Amercia, Europe and Asia [Roth & Kormann, 2000; Lieberei & Reisdorff, 2007; Department of Agriculture, Forestry and Fisheries, 2010; FAOSTAT, 2016].

1.2 Botanic characteristics

Helianthus annuus L is an annual plant. Fully grown, the erect plant is characterized by broad leaves, a strong taproot and a prolific lateral spread of surface roots. In cultivation, sunflowers are usually unbranched and grow to a height of 300 cm or more. They bear one wide flower head, up to a size of 40 cm in diameter, with yellow ray florets at the outside and yellow or maroon disc florets inside. The outer florets are sexually sterile, whereas the inner florets ("disc flowers") mature into fruits. Typically, crop sunflower varieties are self-incompatible. Therefore, pollen movement among plants by insects is important, and bee colonies are often used to increase yields. Each disc flower can produce up to 2000 achenes, single-seeded indehiscent fruit, which are commonly referred to as "sunflower seeds" [Dorrel & Vick, 1997; Department of Agriculture, Forestry and Fisheries, 2010].

1.3 Oil production

The sunflower has one of the shortest growing seasons of the major crops of the world. Early maturing varieties are ready for harvest 90 to 120 days after planting: they require approximately 11 days from planting to emergence, 33 days from emergence to head visibility, and 65 days from head visibility to maturity. Depending on climatic and cultivation conditions, yields can vary from as little as 600 to as much as 3.000 kg/ha [Department of Agriculture, Forestry and Fisheries, 2010; FAOSTAT, 2016].

Ripe fruiting heads are harvested either manually or mechanically. The oil containing seeds are four-sided and flat and are generally 0.8 cm to 1.7 cm long and 0.4 cm to 0.9 cm wide. Additionally, they are wrapped by a hull consisting of a single layer of skin, which can easily be separated [Roth & Kormann, 2000; Lieberei & Reisdorff, 2007].

With an average harvest of 33 million t, sunflower seeds represent 8% of the total global production of oilseeds, following soybean (55%), rapeseed (14%) and cottonseeds (10%). The Ukraine, Russia and Argentina are the main producers of sunflower seeds and sunflower by-products, as well as suppliers to the global market. They account for 52% of the global production of sunflower and 40% of the world exports of sunflower seeds. Therefore, these countries are often referred to as the "Sunflower Triangle" [faostat, 2016].

Sunflower oil is extracted mainly from oil-type sunflower seed varieties and hybrids. Solvent-extracted sunflower meal (SEM), a by-product of the oil extraction process, is used primarily as an ingredient in livestock feed rations. Oil-type sunflower seeds contain 38% to 50% oil and about 20% protein. More than 90% of the sunflower seeds produced are processed into edible oil. Per 100 kg of seeds, about 40 kg of oil, 35 kg of high-protein meal and 20 kg to 25 kg of by-products are produced. Sunflower seeds are processed in industrial oil mills with processing capacities of up to 15.000 t oilseed per day. In order to ensure good storage properties, the water content of the seeds is generally reduced to a maximum of 10%. Sunflower oil manufacture involves different stages: cleaning and grinding the seeds, pressing and extracting the crude oil, and further refining of the oil. For oil extraction, volatile hydrocarbons are used as a solvent [Remmele, 2009; faostat, 2016].

Incoming oil seeds show a high degree of contamination with stones, dirt and metals. To remove any extraneous material, the seeds are passed both over magnets and through differently sized sieves. The dehulled seeds are ground into coarse meal to provide more surface area. Mechanic grooved rollers or hammer mills crush the seeds before the resulting meal is heated to facilitate the extraction of the oil. While this procedure permits the recovery of a higher amount of oil, more impurities are released with the oil, which in turn need to be removed. The heated meal is continuously fed into a screw press, which increases the pressure progressively as the meal passes through a slotted barrel. The remaining oil cake still contains about one third of its original amount of oil. Hexane or a comparable volatile hydrocarbon dissolves the oil out of the oil cake before it is separated by distillation. About 90% of the hydrocarbon in the extracted oil simply evaporates and is collected for reuse. The remaining hydrocarbon is retrieved and recollected by a stripping column [Bockisch, 1993; Remmele, 2009; faostat, 2016].

In order to remove color, odor, and bitterness, the oil is refined at temperatures between 40°C and 85°C in the presence of alkaline substances such as sodium hydroxide or sodium carbonate. Additionally, the oil is treated by water steam or a water/acid mixture and subsequently centrifuged, to eliminate the arising gums (phosphatides, precipitate, and dregs). Oil which is intended to undergo high temperatures ("cooking oil") is usually bleached by filtering it through Fuller's earth, activated carbon, or activated clays that absorb pigmented material from the oil. In contrast, oil that will undergo low temperatures (in the refrigerator, for example) is winterized – rapidly chilled and filtered to remove any waxes. This procedure ensures that the oil will not partially solidify during storage at low temperatures [Bockisch, 1993; Remmele, 2009; faostat, 2016].

During deodorization, steam is passed through the hot oil in a vacuum at between 225°C and 250°C. This process allows any volatile taste and odor components to distil from the oil. After deodorization, citric acid at 1% is added to chelate any trace metals that might promote oxidation within the oil (and hence shorten its shelf life) [Bockisch, 1993; Remmele, 2009; faostat, 2016].

In addition to its final oil content of 3%, the remaining SEM contains hydrocarbon residues. To minimize the risk of explosion during transport and storage, the final content of hydrocarbon must be kept to a maximum 300 mg/kg, which is usually achieved by desolvation at temperatures of up to 108°C [Bockisch, 1993; Münch, 2009].

2 Solvent-extracted sunflower meal

Solvent-extracted sunflower meal (SEM) is the co-product that results from solvent extraction of sunflower oil from the seeds. It contains 400 to 500 g/ kg protein (dry matter basis), making SEM an economically interesting source of protein. For further usage as animal feed, the press cake that results from oil production is milled and commonly pressed into pellets. However, excessive microbial protein degradation in the rumen decreases the efficiency of protein utilization in the small intestine [Bockisch, 1993; González-Pérez et al., 2002; Münch, 2009; Calsamiglia et al., 2010; Weisz et al., 2010; Lomascolo et al., 2012].

Thus, ruminants such as goats or cows benefit from ruminally undegraded protein sources that escape degradation but can still be hydrolyzed (leaving the released amino acids to post-ruminal absorption). Previous attempts aimed at the preparation of high-protein sources include, but are not limited to, the use of polymeric coatings, heat, addition of ionophores, treatment of proteins with chemicals such as formaldehyde or different acids, and the addition of secondary plant metabolites [Antoniewicz et al., 1992; Wu and Papas, 1997; Yu et al., 2002; Patra and Saxena, 2009].

However, some phytogenic additives may induce adverse effects such as decreased feed intake due to sensory effects. The use of formaldehyde and other chemicals has raised safety concerns. Heat treatment facilitates Maillard reactions, which can be poorly controlled and may lead to decreased intestinal digestibility and, consequently, availability of some amino acids. Therefore, alternative approaches for the preparation of feedstuffs with elevated crude protein concentrations are needed. A promising approach may be using naturally occurring compounds in feedstuffs such as chlorogenic acid (CQA).

2.1 Composition

The exact composition of SEM depends on the characteristics of the cultivar and on the method of processing, specifically the degree of dehulling (Table 1.1).

Table 1.1: Composition of SEM (modified from DLG, 1997)

SEM [g · kg^{-1}]	Dry matter (DM) [g · kg DM^{-1}]	Crude ash [g · kg DM^{-1}]	Crude protein [g · kg DM^{-1}]	Crude fat [g · kg DM^{-1}]	Crude fibre [g · kg DM^{-1}]	N-free extractive matter [g · kg DM^{-1}]
Dehulled	910	79	439	20	135	327
Partly dehulled	900	70	379	24	223	304
Unhulled (low hull cultivar)	880	64	324	25	287	300
Unhulled (High hull cultivar)	890	50	237	13	404	296

The residual water content of SEM ranges from 9% to 12%. The crude fiber mainly consists of cellulose, lignin and pentosans. The precise amount depends on the degree of dehulling and ranges from 15% to 40%. Similarily, the protein content varies from 25% to 45%. The content of crude fat figures between 1% and 3%. The dry matter ranges between 5% and 8%. Nitrogen-free extractive matter includes all organic components that are neither crude protein, crude fat, nor crude fiber, such as the soluble components of cellulose, lignin or pentosans or hemicellulose, pectins and carbohydrates. The content of carbohydrates in SEM ranges from 5% to 10% [Gassmann, 1983; DLG, 1997; González-Pérez & Vereijken, 2007].

2.2 Protein

Unprocessed sunflower seeds exhibit a crude protein content of 17% to 22%, depending on the cultivar. Dehulled sunflower seeds contain between 25% and 45% crude protein. The total amount of crude protein includes peptides, free amino acids and further nitrogen-containing compounds that account for up to 13%. Mainly depending on the extraction method, the protein content in SEM varies from 30% to 50% [Gassmann, 1983; González-Pérez & Vereijken, 2007; Weisz et al., 2009].

Furthermore, except for their low lysine content, sunflower proteins match the FAO (Food and Agriculture Organization) reference protein patterns for humans in terms of amino acid composition and are low in antinutritive compounds. In addition to their relatively high nutritive value, sunflower proteins display various interesting technological characteristics comparable to those of soybean and other legume proteins, such as emulsifying or foaming properties [Weisz et al., 2010].

The classification of sunflower proteins is based on either the Osborne fractionation or the Svedberg sedimentation coefficient. The Osborne fractionation classifies the single proteins according to their solubility behavior. In accordance to that, the main part of sunflower protein (50% to 70%) consists of salt-soluble globulines, whereas water-soluble albumins account for only 18% to 35% [Raymond et al., 1995].

Sunflower protein is naturally low in both alkaline-soluble glutelins and alcohol-soluble prolamines. The Svedberg sedimentation coefficient classifies SEM protein into two major fractions of globular proteins, with helianthinin being the predominant one. Helianthinin is an 11S-globuline with a molecular weight of 300 to 350 kDa. Depending on pH, ionic strength, temperature and protein concentration, its hexamer 11S form readily dissociates into its 7S form, which, under extreme conditions, dissociates into its 2–3S form. Additionally, in high concentrations and under moderate alkaline conditions (pH 8.5 to 9), helianthinin aggregates to 15–18S forms. The second major Svedberg fraction includes 2S albumins, which account for 10% to 30% of the total SEM protein content. They are characterized by a molecular weight of 10 to 18 kDa and a high degree of stability towards pH changes as well as under high temperatures [Sabir et al., 1974; Schwenke et al., 1975; Gassmann, 1983; González-Pérez et al., 2004; González-Pérez & Vereijken, 2007].

2.3 Phenolic content

In addition to its high protein content, SEM is also rich in phenolic compounds, with chlorogenic acid (CQA) being the predominant component. Depending on environmental and genetic factors, total phenolics in SEM may range from 1% to 4%, with 5-O-caffeoylquinic acid (chlorogenic acid, CQA) being the predominant compound. Along with its main isomers 3-O-caffeoylquinic acid and 4-O-caffeoylquinic acid, CQA accounts for 62% to 92% of the total polyphenol content in SEM [Sabir et al., 1974; Weisz et al., 2009].

3 Polyphenols

3.1 General overview

Plant phenolic compounds, or polyphenols, are characterized as compounds possessing one or more aromatic rings bearing hydroxyl substituent(s). The name derives from the Ancient Greek word πολύς (polus, meaning "many, much"), and the word phenol, which refers to the chemical structure formed by attaching to an aromatic benzenoid (phenyl) ring, a hydroxyl (-OH) group identical to that found in alcohols (hence the -ol suffix). The term polyphenol appears to have been in use since 1894 [Prigent et al., 2007; Merriam-Webster Thesaurus, 2017].

Polyphenols form a diverse structural class of natural compounds that are synthesized in plant secondary metabolism. They play a role in numerous processes, such as plant growth and reactions to stress and pathogen attack. The main effects of phenolic compounds in plant tissues can be divided into the following categories [Engelhardt & Galensa, 1997; Parr and Bolwell, 2000; Weisz et al., 2010; Le Bourvellec & Renard, 2012]:

- Release and suppression of growth hormones
- UV screens to protect against ionizing radiation and to provide coloration
- Deterrence of herbivores
- Prevention of microbial infections
- Signaling molecules in ripening and other growth processes

The content and profile of phenolic compounds depend on the cultivar and cultivation conditions of the plant. Also, the degree of maturation significantly affects the total phenolic content. The highest concentration of phenolic compounds can be observed in the exterior layer of the single plant parts. Phenolic compounds contribute to the organoleptic properties of plant foods, especially by their astringency, bitter taste and color. They occur naturally in many foods and drinks from plant origin, e.g. fruits, vegetables, coffee, tea, beer, wine and chocolate. The total content of phenolic compounds in foods may change during storage, due to effects induced by light and temperature. Their presence can be easily observed due to the chromophoric groups that some phenolic compounds bear, e.g., the red-purple anthocyanins, or by the brown reaction products of polyphenols when fruits are damaged [Robards et al. 1999; Friedman & Jürgens, 2000; Yabuta et al., 2001; Prigent, 2005].

Owing to their antioxidative properties, phenolic compounds are of great interest for the food industry both nutritionally and technologically. They have been associated with various health-promoting effects such as antioxidative, antibiotic or anti-inflammatory properties, which possibly prevent diseases associated with oxidative stress. Because of their presumed positive effects on human health, polyphenols are increasingly used in functional foods [Prigent, 2005; Kammerer et al., 2007; Weisz et al., 2010].

Phenolic compounds represent a wide range of molecules with a molecular mass from about 100 to 4.000 Da. They can be divided into the following groups [O'Connell & Fox, 2001; Prigent, 2005]:

- the C_6 group, comprising simple phenols and benzoquinones,
- the C_6C_n group, which consists of phenolic acid derivatives and hydroxycinnamic acid derivatives,
- the C_6-C_n-C_6 group, including flavonoids (C_6-C_3-C_6),
- the $(C_6$-$C_3)_n$ group consisting of lignans and lignins, and
- the tannin group, which itself in turn is divided into hydrolyzable tannins, condensed tannins and phlorotannins. The hydrolyzable tannins are formed by gallic acid, or hexahydroxydiphenic acid, esterified to a polyol such as glucose or quinic acid. The condensed tannins mainly consist of the proanthocyanidins. Proanthocyanidins are oligomers and polymers of flavanols, which are members of the flavonoid sub-class. Therefore, some authors classify proanthocyanidins into the same class as their monomeric units, i.e. the C_6-C_n-C_6 class. Among these monomeric units,

(+)-catechin, (-)-epicatechin, (+)-gallocatechin and (-)-epigallocatechin are the most common. Their 2S enantiomers, e.g. (-)-catechin and (+)-epicatechin, may also be present. Phlorotannins are oligomers of phloroglucinol. They are found in brown algae, and in a lower amount also in red algae [Haslam et al., 1989; Van Alstyne et al. 1999, O'Connell and Fox, 2001; Prigent, 2005].

For the studies described in this thesis, mainly 5'-caffeoylquinic acid was considered, a representative of the hydroxycinnamic acids, which belongs to the C_6C_n class.

3.2 Chlorogenic acid in SEM

Approximately 10% of the total phenolic content in SEM consists of chlorogenic acid (CQA) dimers, of which 3,5-di-O-CQA is the predominant one over 4,5-di-O-CQA and 3,4-di-O-CQA. Furthermore, in relatively small quantities, SEM contains non-esterified caffeic acid and, additionally, ferulic and quinic acid derivates. The term chlorogenic acid (CQA; caffeoylquinic acid) usually refers to a family of esters of caffeic acid (a hydroxycinnamic acid). 5'-O-Caffeoylquinic acid (5-CQA; 1',3',4',5'-tetrahydroxycyclohexanecarboxylic acid 3-(3,4)-dihydroxypropenyl-dihydroxyphenyl-1-propenoate; Figure 1.1) is the most common individual compound of the cinnamic acid family and has a mass of 354 Da [Clifford, 2000; Weisz et al., 2009].

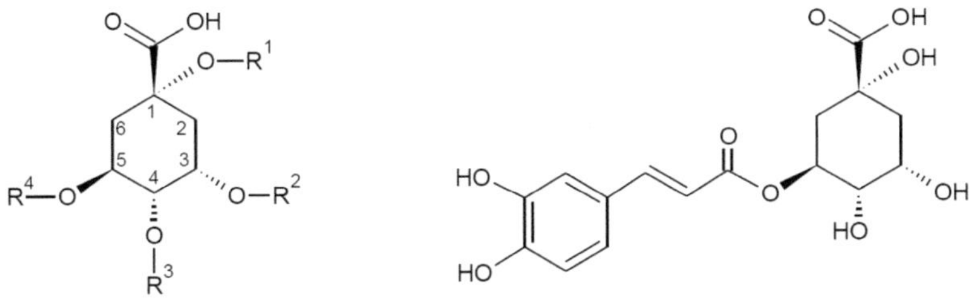

Figure 1.1: Quinic acid (R^1, R^2, R^3, R^4 = H) and 5-O-caffeoylquinicacid [Weisz et al., 2009].

Other major chlorogenic acids are 3'-O-caffeoylquinic acid (3-CQA) and 4'-O-caffeoylquinic acid (4-CQA), both comparatively stable isomers of 5-CQA. Additional chlorogenic acids include diesters such as 3,4-, 3,5- or 4,5-

dicaffeoylquinic acids. CQA is considered to be the main phenolic compound responsible for the browning of fruits. Due to its ubiquitous character, 5-CQA is often used as a model compound [Clifford, 1985; Naczk, 2011; Robards et al., 1999; Clifford, 2000].

4 Polyphenol-protein interactions

As a result of their allelochemical properties, phenolic compounds such as CQA are able to bind to target molecules such as nucleic acids or proteins. In healthy plant organisms, the single constituents are located in different compartments. Thus, interactions between polyphenols and proteins or peptides only occur upon damage to these structures by mechanical pressure, for instance. In nature, the principle purpose of such interactions is to fend off herbivores or rivalling plants. Accordingly, polyphenol-protein interactions are often regarded as defense mechanisms of plants [Le Bourvellec & Renard, 2012].

In plant food, polyphenol-protein interactions contribute to the organoleptic properties, particularly to astringency. Astringency, a characteristic of many beverages such as wine or tea, is associated specifically with the interaction of salivary proteins with polyphenols. This interaction results in insoluble aggregates that precipitate and interfere with palate lubrification [Haslam, 1996; Baxter et al., 1997; Le Bourvellec & Renard, 2012].

Another possible organoleptic effect of polyphenol-protein interactions is color development. Interactions between polyphenol-derived quinones and free amino acids or protein side chains often result in colored pigments. A prominent example of this effect is green discoloration during potato cooking [Friedman & Jürgens, 2000; Namiki et al., 2001, Yabuta et al., 2001].

In addition, polyphenol-protein interactions may lead to both soluble and insoluble complexes, which may result in colloidal haze. Interactions between polyphenols and proteins can also limit beverage quality due to visible haze formation [Friedman & Jürgens, 2000; Siebert et al.,1996; Siebert, 1999; Siebert & Lynn, 2000].

In the digestive tract, larger polyphenols such as tannins are likely to have a negative impact on protein digestibility, as has been shown both *in vivo* and *in vitro*. Complexation between polyphenols and proteolytic and glycolytic enzymes may result in the inhibition of the latter, thereby reducing overall digestion of the food. Moreover, polyphenols and food proteins are likely to form soluble or insoluble complexes, thereby reducing their bioavailability. This is not a problem in modern Western diets, which are typically rich in proteins, but may have negative consequences in more restricted conditions [Friedman & Jürgens, 2000; O'Connell & Fox, 2001; Cheynier, 2005; He et al., 2007; Le Bourvellec & Renard, 2012].

However, polyphenol-protein interactions may also have positive effects on functional properties of proteins, for instance, by improving foam formation.These interactions are exploited in food processing operations like clarification and fining treatments: adding proteins such as caseine to beverages results in the precipitation of tannins, thus improving haze stability and taste [Siebert et al., 1996; Siebert, 1999; Rossetti et al., 2008; Le Bourvellec& Renard, 2012].

Interactions between polyphenols and proteins occur during the entire course of plant-based food processing due to cell destruction and the resulting exposure to oxygen from the ambient atmosphere. At the core of these interactions lies the ability of polyphenols to form bonds, both non-covalent and covalent, with proteins leading to transient and/or irreversible complexes, respectively. Both covalent and noncovalent bindings are likely to take place simultaneously. In food products, the interactions responsible for the effects on functional properties of proteins are generally due to non-covalent reactions with oligomeric phenolics and to covalent interactions with both monomeric and oligomeric phenolics [Prigent et al., 2003; Prigent 2005; Prigent et al., 2007].

Initially, association and dissociation between proteins and polyphenols is a surface phenomenon. The stability of complexes depends on the molecular size of the interacting partners, their concentrations and the external conditions in which the interaction takes place. Polyphenols of higher molecular size show a greater tendency to form stable complexes with proteins [Spencer et al., 1988; Haslam, 1996; De Freitas & Mateus, 2001].

In this respect, it is important to realize that proteins and polyphenols vastly differ in size. The average size of a polyphenol is ~500 Da, whereas a typical size of a protein is 30.000 Da. To produce stable complexes, the ratio of polyphenol to protein needs be sufficiently high to form multiple non-covalent bonds. Additionally, the strength of complexes is influenced by polyphenol hydrophobicity. At low polyphenol concentrations, their attachment to proteins does not change the hydrophilic character of complexes and the complexes remain soluble. However, with increasing concentration of polyphenols, the protein-polyphenol complexes become more hydrophobic by encouraging interactions between non-polar residues of polyphenols and proteins. The intramolecular hydrophobic interactions between different protein- polyphenol complexes are the driving force toward aggregation and precipitation as large, insoluble aggregates [Spencer et al., 1988; Naczk et al., 2011].

Irrespective of the exact nature of interaction, two types of complexation mechanisms are distinguished: the monodendate mechanism and the multidendate mechanism (shown in figure 1.2). If the polyphenol interacts with only one protein site, the mechanism is monodentate. In the event of a high polyphenol-to-protein ratio, polyphenols form a layer around a protein molecule, thereby covering its surface. As a consequence, the protein becomes less hydrophobic, which may result in aggregation and precipitation. The multidentate mechanism applies only to polyphenols with sufficient size to interact with more than one site, thus being able to form cross-links between proteins. Aggregation and precipitation can also emerge, but at a much lower polyphenol-to-protein ratio compared to the monodendate mechanism [Haslam et al., 1989; Charlton et al., 2002; Prigent, 2005].

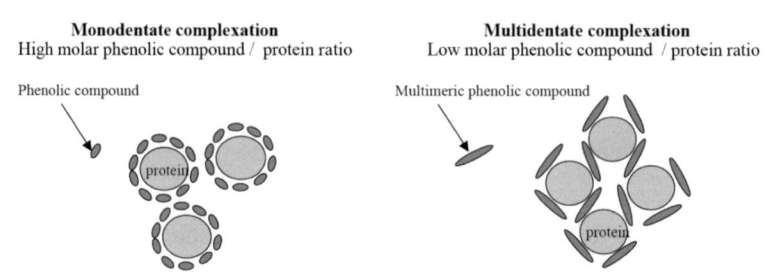

Figure 1.2: Monodentate and multidentate mechanisms [adapted from Haslam, 1989]

Proanthocyanidins, for example, decrease protein solubility at considerably lower ratios than monomeric polyphenols such as CQA. Accordingly, the precipitation not only depends on the steric hindrance, but, next to molecular weight of the polyphenol, on its conformational mobility and flexibility as well as its solubility and the polarity of both the polyphenol and the protein. Therefore, the nature of the protein chains is of particular importance [Prigent, 2005; Le Bourvellec & Renard, 2012].

The size and solubility of protein polyphenol complexes are temperature-dependant; elevated temperatures facilitate tendency to hydrophobic interactions by unfolding the proteins chains and exposing hydrophobic amino acids (hydrophobic effect). Thus, the size and number of protein-polyphenol complexation will increase with a rise in temperature [Spencer et al., 1988; Naczk et al., 2011].

4.1 Non-covalent interactions

Under acidic conditions, non-covalent interactions are dominant, such as the formation of hydrogen bonds and hydrophobic interactions. Hydrogen bonds emerge from the interaction between the electronegative atoms of amino- and hydroxyl groups, and positively charged hydrogen atoms from any other amino- or hydroxylgroup within the molecule or from another molecule. Depending on the polyphenol structure and the degree of hydroxylation, the interaction may produce single or multiple hydrogen bonds, which is decisive for the strength of the resulting complexes. Hydrogen bonds between neighboring protein chains may result in bridges that crosslink proteins into aggregates [Haslam, 1974; Haslam 1996].

In addition to hydrogen bonds, proteins and polyphenols may interact via hydrophobic, non-polar aromatic rings of polyphenols and aromatic amino acids, e.g. proline, phenylalanine, tyrosine, tryptophan, histidine [Charlton et al., 2002; Siebert, 1999].

Polyphenol interactions with proline-rich proteins show a particularly high degree of hydrophobic interaction, which in turn stabilzes the respective aggregates [Baxter et al., 1997; Siebert et al., 1996].

Protein-polyphenol complexes usually emerge from multiple cooperative hydrophobic and hydrogen binding and may lead to colloidal size aggregates, especially in aqueous solutions. As a consequence, a complete separation of CQA in SEM via polar organic, aqueous or aqueous-organic solvents can be difficult [Sabir et al., 1974; Prigent et al., 2007].

4.2 Covalent interactions

Under alkaline conditions or in presence of the enzyme polyphenoloxidase, irreversible interactions between polyphenols and proteins dominate. The underlying reaction mechanism is the formation of quinones: *o*-diphenolic compounds can be oxidized to *o*-quinones. These quinones can be formed enzymatically or non-enzymatically. The covalent structures result from oxidation and nucleophilic addition processes. Heat treatment enhances the formation of quinones to a certain degree, but may also result in the permanent destruction of both the phenolic structures and the proteins [O'Connell & Fox, 1999; Haslam, 1996; Friedman & Jürgens, 2000; Prigent et al., 2003, Le Bourvellec & Renard, 2012].

4.2.1 Enzymatic Oxidation

Polyphenoloxidases (PPO; EC 1.14.18.1) are divided into catechol oxidase (tyrosinase) and laccase [Mayer and Harel, 1979; Nicolas et al., 1994; Le Bourvellec & Renard, 2012].

In the presence of molecular oxygen, they catalyze the oxidation of phenolic substrates to their relating o-quinone via a 2-electon-transfer, as shown in Figure 1.3.

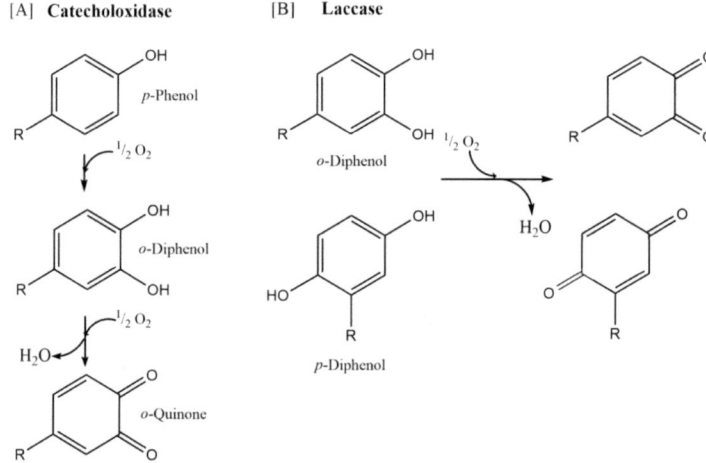

Figure 1.3: Enzymatic formation of o-quinones (A) with catecholoxidase, (B) with laccase [modified from Nicolas et al., 1994]

Tyrosinase catalyzes two different kinds of reaction
1. Hydroxylation of monophenols into the relating o-diphenols
2. Formation of o-quinones from o-diphenols

Laccase, in contrast to tyrosinase, does not catalyze the hydroxylation of monophenols. Instead, it initiates the oxidation of a broad spectrum of substrates, including p- und o-diphenols. When o-diphenols are in eccess, a comproportionation takes place, which results in the formation of o-semi-radicals [Matheis & Whitaker, 1984; Le Bourvellec & Renard, 2012].

In addition to binding the nucleophilic amino acids groups, *ortho*-semi-quinone radicals may undergo radical-radical reactions between themselves, generating oligomers and polyphenol polymers that are characterized by a somewhat brown color [Cilliers & Singleton, 1991; Hotta et al., 2001; Namiki et al., 2001; Bors et al., 2004].

4.2.2 Alkaline oxidation

Due to the extraction process, SEM does not contain any active polyphenoloxidase. However, under alkaline conditions, the phenolic hydroxyl groups of the *o*-diphenols are deprotonated, leading to the oxidation of the molecules. As shown in Figure 1.4, the resulting semi-quinones immediately react to *o*-quinones. These, in turn, readily form CQA dimers or react with primary amino groups or terminal amino acids in proteins and peptides [Cilliers & Singleton, 1991; Prigent et al., 2007; De Leonardis et al., 2010].

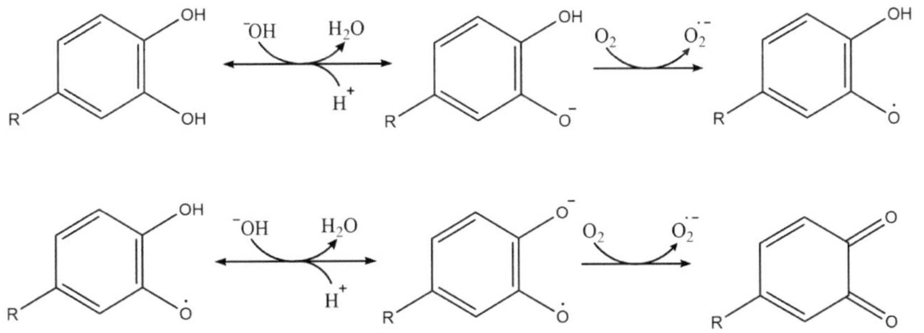

Figure 1.4: Alkali-induced formation of *o*-quinones and *o*-semi-quinones [modified from Le Bourvellec & Renard, 2012]

4.2.3 Covalent interactions between quinones and proteins or other nucleophilic molecules

In case of CQA, covalent reaction products usually consist of distinctly conjugated dimers and trimers that absorb in the visible wavelength range.

As a result of a nucleophilic attack by amino or thiol groups, colored reaction products may emerge. This is not the case for all polyphenols. Free caffeic acid, ferulic acid, isoferulic acid and *p*-coumaric acid do not develop any colored products, and neither do dihydrocaffeates. Deriving from that, the presence of an *o*-diphenol structure as well as a double bond and a carbonyl group appears likely [Sabir et al., 1973; Cilliers & Singleton, 1991; Yabuta et al., 2001; Prigent et al., 2008; Schilling et al., 2008].

5 Trihydroxy benzacridine derivates

Incubation of CQA and separate amino acids under alkaline conditions or in the presence of PPO may result in green discoloration of the reaction mixture due to oxidation. Similar results were observed in alkaline treated model solutions containing both ethyl-caffeate and *n*-butyl-amine, whose characteristics are analogous to the lysine side chain. Yabuta et al. (2001) identified the responsible green pigment as an oxidized trihydroxy benzacridine derivative, which emerged from condensation of two ethyl caffeates and a primary amino compound. As shown in Figure 1.5, the underlying reaction mechanism is assumed to be a dimerization, subsequent to the binding of the isoprenyl groups of two reactive semi-quinone forms of ethyl caffeate. The dimerization is followed by further oxidation and a Michael addition between the dimer and an amino compound. The subsequent nucleophilic cyclization results in the separation of a single hydrogen atom, thus yielding the final trihydroxy benzacridine derivative. This reaction mechanism can also be observed in model solutions containing CQA and *N*-BOC-lysine [Namiki et al., 2001; Yabuta et al., 2001; Prigent, 2005; Schilling et al., 2008].

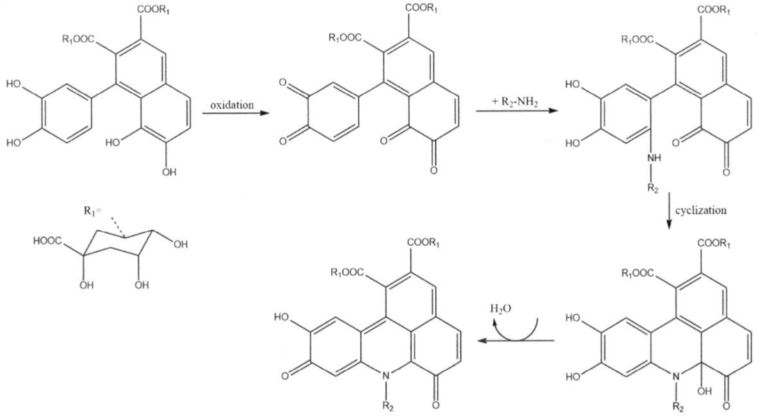

Figure 1.5: Proposed reaction mechanism between CQA and individual amino acids. R1 = Quinic acid. R2 = amino side chain [modified from Namiki et al., 2001; Yabuta et al., 2001]

6 Aim

Consumption of fruits and vegetables is no longer just a matter of taste; it is also a matter of health. Plant based food is an important source of primary nutrients such as proteins and secondary metabolites such as phenolic compounds. Polyphenols are associated with positive health effects, e.g. in the prevention of cardiovascular and neurodegenerative diseases. During the last decade, great progress has been made in the recovery of functional foods from natural sources. Many chemical and biological studies focused on the identification of sources of bioactive compounds, elucidation of the parameters associated with their bioaccessibility and bioavailability, as well as on deciphering their mechanisms of action once they are absorbed in the body. However, sufficient information about interactions between phenolic compounds and other nutrients is still lacking. For example, interactions with isolated, proline-rich proteins, have been extensively studied, whereas polyphenol interactions with food proteins are still poorly understood. Usually, the single constituents are separate in different plant compartments. During food processing, these structures are disintegrated, for instance, due to mechanical impact. As a result, polyphenols and food protein are likely to form soluble or insoluble complexes. These can have positive effects on functional properties of proteins, which are exploited in food processing operations, e.g. by improving foam formation. Unfortunately, also negative effects may occur. A prominent example for this is green discoloration during potato cooking. For human nutrition, insoluble polyphenol-protein complexes may result in a reduced bioavailability, both of protein and polyphenols, or even a decreased overall digestion of the food. However, what is a disadvantage in human nutrition might well be an advantage in animal nutrition: protein is an important component in animal feed. Especially ruminants such as goats or cows need a comparatively high amount of it, as feed protein is target to mibrobial degradation in their rumen. As a consequence, the released amino acids are further degraded microbially, and thus, their bioavailability is decreased. This study aimed at evaluating the potential use of polyphenol-protein complexes in animal nutrition. Special attention was given to ruminally undegraded protein that escapes ruminal degradation and in turn can be hydrolyzed and absorbed post-ruminally. The subsequent examination of corresponding model systems aimed at understanding the underlying interreaction mechanism between chlorogenic acid and single amino acids as well as the identification of resulting interaction products.

References

Antoniewicz, A.; van Vuuren, A.; van der Koelen, C.; Kosmala, I. (1992). Intestinal digestibility of rumen undegraded protein of formaldehyde-treated feedstuffs measured by mobile bag and in vitro technique. Anim Feed Sci Technol 39, 111–124.

Asch, D. L. (1993). Common sunflower (*Helianthus annuus* L.): The pathway toward its domestication. Proceedings of the 58th Annual Meeting 1993. Society of American Archaeology, St. Louis, MO, USA, 1–15

Bassil, D.; Makris, D.P.; Kefalas, P. (2005). Oxidation of caffeic acid in the presence of L-cysteine: Isolation of 2-S-cystinylcaffeic acid and evaluation of its antioxidant properties. Food Res Int 38, 395–402.

Baxter N.J.; Terence H.; Lilley T.H.; Haslam E.; Michael P.; Williamson M.P. (1997). Multiple interactions between polyphenols and a salivary proline-rich protein repeat result in complexation and precipitation. Biochemistry 36, 5566–5577.

Bockisch, M. (1993). Nahrungsfette und -öle. Stuttgart: Ulmer Verlag.

Bors W.; Heller, W.; Michel C.; Saran M. (1990). Flavonoids as antioxidants: determination of radical scavenging efficiencies. Methods Enzymol 186, 343–355.

Bors W.; Michel C; Stettmaier K.; Lu Y.; Foo L.Y. (2004). Antioxidant mechanism of polyphenolic caffeic acid oligomers, constituent of *Salvia officinalis*. Biol Res 37, 301–311.

Calsamiglia, S.; Ferret, A.; Reynolds, C.; Kristensen, N.; van Vuuren, A. (2010). Strategies for optimizing nitrogen use by ruminants. Animal 4, 1184–1196.

Charlton A.J.; Baxter N.J.; Lokman Khan M.; Moir A.J.G.; Haslam E.; Davies A.P.; Williamson M.P. (2002). Polyphenol/peptide binding and precipitation. J Agric Food Chem 50, 1593–1601.

Cheynier V. (2005). Polyphenols in foods are more complex than often thought. Am J Clin Nutr 81, 223–229.

Cilliers, J.J.L.; Singleton, V.L. (1991). Characterization of the products of nonenzymic autoxidative phenolic reactions in a caffeic acid model system. J Agric Food Chem 39, 1298–1303.

Clifford M.N. (1985). Chemical and physical aspects of green coffee and coffee products. In: Coffee: Botany, Biochemistry and Production of Beans and Beverage. Croom Helm, London, 305–374.

Clifford, M.N. (2000). Anthocyanins – nature, occurrence and dietary burden. J Sci Food Agric 80, 1063–1072.

Clifford, M.N.; Johnston, K.L.; Knight, S.; Kuhnert, N. (2003): Hierarchical scheme for LC-MSn identification of chlorogenic acids. J Agric Food Chem 51, 2900–2911.

De Freitas V.; Mateus N. (2001). Structural features of procyanidin interactions with salivary proteins. J Agric Food Chem 49, 940–945.

Department of Agriculture, Forestry and Fisheries. Sunflower production guideline. 2010. http://www.nda.agric.za/docs/Brochures/prodGuideSunflower.pdf accessed 22.02.2017

DLG – Deutsche Landwirtschafts-Gesellschaft (1997). Futterwerttabellen – Wiederkäuer. 7. Auflage, Frankfurt am Main: DLG-Verlag.

Dorrel, D. G.; Vick, B. A. (1997). Properties and processing of oilseed sunflower. In: Schneiter, A.A.: Sunflower Technology and Production. Madison, Wisconsin: American Society of Agronomy, 709–74.

Engelhardt, U.H.; Galensa, R. (1997). Analytik und Bedeutung von Polyphenolen in Lebensmitteln. Analytiker-Taschenbuch Bd. 15, Springer-Verlag Heidelberg, 149–178.

FAOSTAT. Food and Agricultural commodities production (2017).
URL: http://faostat.fao.org/site/567/DesktopDefault.aspx?PageID=567#ancor

Friedman, M.; Jürgens, H.S. (2000). Effect of pH on the stability of plant phenolic compounds. J Agric Food Chem 48, 2101–2110.

Gaßmann, B. (1983). Preparation and application of vegetable proteins, especially proteins from sunflower seed, for human consumption. An approach. Die Nahrung 27, 351–369.

González-Pérez, S.; Merck, K. B.; Vereijken, J. M.; Van Koningsveld, G. A.; Gruppen, H.; Voragen, A.G. (2002). Isolation and characterization of undenatured chlorogenic acid free sunflower (*Helianthus annuus*) proteins. J Agric Food Chem 50, 1713–1719.

González-Pérez, S.; Vereijken, J. M. (2007). Sunflower proteins: Overview of their physicochemical, structural and functional properties. J Sci Food Agric 87, 2173–2191.

Haslam E. (1974). Polyphenol-protein interactions. Biochem J 139, 285–288.

Haslam E.; Phillipson, J.D.; Ayres, D.C.; Baxter, H., Eds. (1989). Plant polyphenols: vegetable tannins revisited. Cambridge University Press: Cambridge, 167–192.

Haslam E. (1996). Natural polyphenols (vegetable tannins) as drugs: possible modes of action. J Nat Prod 39, 205–215.

He Q.; Lv Y.; Yao K. (2007). Effects of tea polyphenols on the activities of α-amylase, pepsin, trypsin and lipase. Food Chem 101, 1178–1182.

Heiser, C.B. (1951). The sunflower among the North American Indians. Proc American Philos Soc 95: 432–448.

Heiser, C.B.; Smith, D.M; Clevenger, S.B.; Martin Jr., W.C. (1969). The North American sunflowers (*Helianthus*). Memoirs of the Torrey Bot Club 22: 1–218.

Hotta H.; Sakamoto H.; Nagano S.; Osakai T.; Tsujino J. (2001). Unusually large number of electrons for the oxidation of polyphenolic antioxidants. Biochim Biophys Acta 1526, 159–167.

Kammerer, D.R.; Saleh, Z.S.; Carle, R.; Stanley, R.A. (2007). Adsorptive recovery of phenolic compounds from apple juice. Eur Food Res Technol 224, 605–613.

Le Bourvellec, C.; Renard, C.M. (2012). Interactions between polyphenols and macromolecules: Quantification methods and mechanisms. Crit Rev Food Sci and Nutr 52, 213–248.

Lieberei, R.; Reisdorff, C. (2007) Nutzpflanzenkunde. 7., überarb. u. erw. Auflage, Stuttgart: Thieme Verlag.

Matheis, G.; Whitaker, J. (1984). Modifications of proteins by polyphenoloxidase and peroxidase and their products. J Food Biochem 8, 137–162.

Mayer, A. M.; Harel, E. (1979). Polyphenoloxidases in plants. Phytochemistry 18, 193–215.

Münch, E. W. (2009). Ölgewinnung und -veredelung in zentralen Ölmühlen. In: Matthäus, B.; Münch, E. W.: Warenkunde Ölpflanzen/ Pflanzenöle. Clenze: Agrimedia, 121–251.

Naczk M.; Towsend M.; Zadernowski R.; Shahidi F. (2011); Protein-binding potential of phenolics of mangosteen fruit (Garcinia mangostana). Food Chem 128, 292–298.

Namiki, M.; Yabuta, G.; Koizumi, Y.; Yano, M. (2001): Development of free radical products during the greening reaction of caffeic acid esters (or chlorogenic acid) and a primary amino compound. Biosci Biotech Biochem 65, 2131–2136.

Nicolas, J.J.; Richard-Forget F.C.; Goupy P.M.; Amiot M.J.; Aubert S.Y. (1994). Enzymatic browning reactions in apple and apple products. Crit Rev Food Sci Nutr 34, 109–57.

O'Connell, J.E.; Fox, P.F. (2001). Significance and applications of phenolic compounds in the production and quality of milk and dairy products: a review. Int Dairy J 11, 103–120.

Parr, A.; Bolwell, G.P. (2000). Phenols in the plant and in man. The potential for possible nutritional enhancement of the diet by modifying the phenols content or profile. J Sci Food Agric 80, 985–1012.

Prigent, S.V.; Visser, A.J.; Jong, G.A. de; Gruppen, H.; van Koningsveld, G.; van Koningsveld, Gerrit A. et al. (2003): Effects of non-covalent interactions with 5-O-caffeoylquinic acid (chlorogenic acid) on the heat denaturation and solubility of globular proteins. J Agric Food Chem 51, 5088–5095.

Prigent, S.V. (2005). Interactions of phenolic compounds with globular proteins and their effects on food-related functional properties. Dissertation. Wageningen University, Netherlands.

Prigent, S.V.; Voragen, A.G.; van Koningsveld, G.; Gruppen, H.; Visser, A. (2007). Covalent interactions between proteins and oxidation products of caffeoylquinic acid (chlorogenic acid). J Sci Food Agric 87, 2502–2510.

Prigent, S.V.; Li, Feng; van Koningsveld, G.; Voragen, A.; Visser, A.; Gruppen, H. (2008). Covalent interactions between amino acid side chains and oxidation products of caffeoylquinic acid (chlorogenic acid). J Sci Food Agric 88,1748–1754.

Remmele, E.: Ölgewinnung und -veredelung in zentralen Ölmühlen. In: Matthäus, B.; Münch, E. W.: Warenkunde Ölpflanzen/ Pflanzenöle. Clenze: Agrimedia, 121–251.

Richard, F.; Goupy, P.; Nicolas, J.; Lacombe, J.; Pavia, A. (1991). Cysteine as an inhibitor of enzymic browning. 1. Isolation and characterization of addition compounds formed during oxidation of phenolics by apple polyphenoloxidase. J Agric and Food Chem 39, 841–847.

Robards, K.; Prenzler, P.; Tucker, G.; Swatsitang, P.; Glover, W. (1999): Phenolic compounds and their role in oxidative processes in fruits. Food Chem 23, 401–436.

Rossetti, D.; Yakubov, G. E.; Stokes, J. R.; Fuller, G. G. (2008). Interaction of human whole saliva and astringent dietary compounds investigated by interfacial shear rheology. Food Hydrocoll 22, 1068–1078.

Roth L., Kormann. K, Ölpflanzen, Pflanzenöle Fette, Wachse, Fettsäuren. Botanik, Inhaltsstoffe, Analytik. Verlag: ecomed. ISBN 13: 9783609687001

Sabir, M.A., F.W. Sosulski, J.A. Kernan (1974). Phenolic Constituents in Sunflower Flour. J Agric Food Chem 22, 124–129.

Sabir, M.A., F.W. Sosulski, S.L. MacKenzie (1973). Gel Chromatography of Sunflower Proteins. J Agric Food Chem 21, 232–239.

Sarker, D.K.; Wilde, P.J.; Clark, D.C. (1995). Control of surfactant-induced destabilization of foams through polyphenol-mediated protein-protein interactions. J Agric Food Chem, 43, 295–300.

Schilling E., Linder C., Noyes R., Rieseberg L. (1998). Phylogenetic relationships in *Helianthus* Asteraceae based on nuclear ribosomal DANN internal transcribed spacer region sequence data. Syst Bot 81, 248–254.

Schwenke, K. D., M. Schultz, K. J. Linow (1975). Über Samenproteine. 5. Mitt. Dissoziationsverhalten des 11-S-Globulins aus Sonnenblumensamen. Die Nahrung 19, 425–432.

Seiler, G.; Rieseberg L. (1997). Systematics, origins and germplasm resources of the wild and domesticated sunflower. In: Sun*flower Science and Technology* 21–66. American Society of Agronomy, Crop Science Society of America, Madison, WI, USA.

Siebert K.J. (1999). Effects of protein-polyphenol interactions on beverage haze, stabilization and analysis. J Agric Food Chem 47, 353–362.

Siebert K.J., Troukhanova N.V., Lynn P.Y. (1996). Nature of polyphenols-protein interactions. J Agric Food Chem 44, 80–85.

Siebert, K.J.; Lynn, P.Y. (2000). Effect of protein-polyphenol ratio on the size of haze particles. J Am Soc Brew Chem 58, S. 117–123.

Smith B.D. (2011). The cultural context of plant domestication in Eastern North America. Curr Anthropol, 52, 417–418

Spencer C.M.; Cai Y.; Martin R.; Gaffney S.H.; Goulding P.N. (1988). Polyphenol complexation – some thoughts and observations. Phytochem 27, 2397–2409.

Van Alstyne, K.; McCarthy, J.; Hustead, C.; Kearns, L. (1999). Phlorotannin allocation among tissues of Northeastern Pacific kelps and rockweeds. J Phycol 35, 483–492.

Weisz, G. M.; Kammerer, D. R.; Carle, R. (2009). Identification and quantification of phenolic compounds from sunflower (*Helianthus annuus* L.) kernels and shells by HPLC-DAD/ESI-MSn. Food Chem 115, 758–765.

Weisz, G. M.; Schneider, L.; Schweiggert, U.; Kammerer, D. R.; Carle, R. (2010). Sustainable sunflower processing – I. Development of a process for the adsorptive decolorization of sunflower [*Helianthus annuus* L.] protein extracts. Inn Food Sci Emerg Technol 11, 733–741.

Wu, S.; Papas, A. (1997): Rumen-stable delivery systems. Adv Drug Delivery Rev 28, 323–334.

Yabuta, G.; Koizumi, Y.; Namiki, K.; Hida, M.; Namiki, M. (2001). Structure of green pigment formed by the reaction of caffeic acid esters (or chlorogenic acid) with a primary amino compound. Biosci Biotech Biochem 10, 2121–2130.

Yu, P.; Goelema, J.; Leury, B.; Tamminga, S.; Egan, A. (2002): Using the DVE/OEB model to determine optimal conditions of pressure toasting on horse beans (*Vicia faba*) for the dairy feed industry. Anim Feed Sci Technol 99, 141–176.

Chapter 2: Protection of protein from ruminal degradation by alkali-induced oxidation of chlorogenic acid in sunflower meal[*]

Lactating ruminants require an adequate supply of absorbable amino acids for the synthesis of milk protein from two sources, i.e., crude protein (CP) synthesized microbially in the rumen and ruminally undegraded CP (RUP) from feed which can both be digested in the small intestine. Several chemical and physical methods have been identified as being effective in increasing the proportion of RUP of total CP of a feedstuff, yet there is a continuing need for developing and establishing methods which protect feed protein from ruminal degradation with acceptable expenditure of labour and other costs. The objective of this study was to identify and quantify effects of and interactions between chlorogenic acid and protein in solvent-extracted sunflower meal (SEM) as induced by alkali treatment. Response surface methodology was employed to investigate the influence of pH, reaction time and drying temperature on the resulting SEM and, subsequently, its protein value for ruminants estimated from laboratory values. For this purpose, alkali-treated SEM was subjected to a fractionation of feed CP according to the Cornell net carbohydrate and protein system as a basis for estimating RUP at different assumed ruminal passage rates (K_p). To estimate the intestinal digestibility of the treated SEM and its RUP, a three-step enzymatic *in vitro* procedure was applied. Alkaline treatment of SEM increased RUP values with factors ranging from approximately 3 (K_p=0.08/h) to 12 (K_p=0.02/h). Furthermore, the intestinal digestibility of the alkali-treated SEM was enhanced by approximately 10% compared to untreated SEM. Increasing pH and reaction time led to both increasing RUP values and intestinal digestibility. In conclusion, a targeted alkaline treatment of naturally occurring compounds in feedstuffs might be a

---

[*] This chapter has been published:
V. Bongartz; C. Böttger; N. Wilhelmy; N. Schulze-Kaysers; K.-H. Südekum; A. Schieber (2018): Protection of protein from ruminal degradation by alkali-induced oxidation of chlorogenic acid in sunflower meal. J Anim Physiol Anim Nutr 102 (1), 209 –215.

promising approach to provide high-RUP feeds for ruminants which, at the same time, have improved intestinal digestibility values.

Keywords: interactions, polyphenols, protected protein, protein degradation, rumen

1 Introduction

The efficient utilization of dietary crude protein (CP) is crucial in ruminant nutrition, both for an optimal supply of nitrogen (N) to ruminal microbes and amino acids to the animal, and for reducing environmental pollution from ruminant husbandry. However, excessive microbial CP degradation in the rumen decreases the efficiency of protein utilization in the small intestine [Calsamiglia et al., 2010].

Thus, ruminants would benefit from ruminally undegraded CP (RUP) sources that escape degradation but can be hydrolyzed and the released amino acids be absorbed post-ruminally. Previous attempts aiming at the preparation of high-RUP sources include, but are not limited to, the use of polymeric coatings, heat, addition of ionophores, treatment of proteins with chemicals such as formaldehyde or different acids, and the addition of secondary plant metabolites [Antoniewicz et al., 1992; Wu and Papas, 1997; Yu et al., 2002; Patra and Saxena, 2009].

However, some phytogenic additives may cause adverse effects such as decreased feed intake due to sensory effects. The use of formaldehyde and other chemicals has raised safety concerns. Heat treatment facilitates Maillard reactions, which can be poorly controlled and may lead to decreased intestinal digestibility and, finally, bioavailability of some amino acids. Therefore, alternative approaches for the preparation of feedstuffs with elevated RUP concentrations are needed. A promising approach may be using naturally occurring compounds in feedstuffs such as chlorogenic acid (CQA) in solvent-extracted sunflower meal (SEM). Sunflower meal is the co-product resulting from solvent extraction of sunflower oil from the seeds and contains 400–500 g/kg CP (dry matter basis) [Lomascolo et al., 2012].

Therefore, SEM is an economically interesting source of nutrients [González-Pérez et al., 2002].

The highest protein yields are usually obtained by alkali extraction, whereas the presence of CQA in the meal gives rise to the formation of chlorogenic *o*-

quinones due to oxidation. The CQA quinones are electron acceptors that readily react with nucleophiles, such as thiols and amino groups in proteins [Rawel et al., 2005; Weisz et al., 2010].

Interactions between CQA quinones and SEM protein may lead to the formation of covalent reaction products, thereby protecting SEM protein from ruminal degradation. It is not known if the covalent reaction products are persisting in the abomasum and small intestine. In the present study, SEM was subjected to alkaline treatment to investigate the influence of reactions between SEM protein and CQA quinones on ruminal degradation in order to enhance the content of RUP. For this purpose, response surface methodology was applied with pH, reaction time and drying temperature of the resulting SEM as independent numerical variables. This approach allows the identification of variables significantly influencing the reaction between CQA and SEM protein and of interactions between these variables as well as a multivariate optimization of the process. The resulting SEM was subjected to fractionation of feed CP according to the Cornell net carbohydrate and protein system (CNCPS) as a basis for estimating RUP. To estimate the intestinal digestibility of the treated SEM, a three-step *in vitro* method was applied.

2 Materials and methods

2.1 Alkali treatment of sunflower meal

Pelleted SEM was purchased from Cefetra B.V. (Rotterdam, The Netherlands). The pellets were ground through a 1 mm screen using the centrifugal mill ZM 100 (Retsch, Haan, Germany) at 14 000 rpm.

Of the ground SEM, 18 subsamples (experimental design given in subsection 'Data treatment and statistical analysis') were subjected to alkaline treatment with variations in duration, pH and temperature of subsequent drying. Alkaline treatment was carried out for 2 h, 10 h and 18 h, respectively, in a beaker using a Stuart orbital shaker at 300 rpm (Bibby Scientific Limited, Stone, Staffordshire, UK). The SEM and deionized water were mixed at a meal-to-water ratio of 0.125 g/ml. The suspensions were adjusted from their original pH 6.8 to pH 7, 8.5 and 10 with 1M NaOH initially, and again – to the respective pH values – after 10 and further 20 minutes. To terminate the reaction, the suspensions were adjusted to pH 6.8 using 1M HCl.

To characterize any discoloration that emerged from the sample treatment, a photometer was used (model Genesys 6, Thermo Fischer Scientific, Schwerte, Germany). To obtain a dry meal for subsequent examinations, all samples were dried at 70 °C, 85 °C or 100 °C for 24 h – 48 h, using a drying cabinet (model 5042E, Heraeus, Hanau, Germany). The resulting samples had a dry matter content of approximately 92% (± 0.1%; relative to the dry matter of the untreated SEM). Untreated SEM served as a control sample.

2.2 Chemical analysis

Proximate analyses on the untreated SEM (one batch, all analyses done at least in duplicate) were carried out according to the Association of German Agricultural Analytic and Research Institutes (VDLUFA, 2012) as indicated by method numbers. The dry matter content was determined by oven-drying of a subsample at 105°C (3.1). Ash and crude lipids were analyzed using methods 8.1 and 5.1.1, respectively. Crude protein was determined by Kjeldahl method

(4.1.1) using a Vapodest 50 s carousel (Gerhardt, Königswinter, Germany) for automated distillation and titration. Neutral detergent fibre (aNDFom; 6.5.1; assayed with heat stable amylase and without sodium sulfite), acid detergent fibre (ADFom; 6.5.2) and acid detergent lignin (ADL; 6.5.3) were analyzed using Ankom^{2000} Fiber analyzer (Ankom Technology, Macedon, USA). Analysis of aNDFom and ADFom was done separately and values were expressed exclusive of residual ash.

2.3 Crude protein fractionation and estimation of ruminally undegraded crude protein

Crude protein in both untreated and treated SEM was categorized into five subfractions (i.e., A, B1, B2, B3, C) based on the CNCPS [Sniffen et al., 1992). For this purpose, true protein (TP), buffer insoluble CP, neutral-detergent insoluble CP (NDICP), and acid-detergent insoluble CP (ADICP) were specified according to standardizations of Licitra et al. (1996) using Kjeldahl digestion to determine N in samples and residues (method 4.1.1, VDLUFA, 2012). All analyses were carried out in triplicate. In short, fraction A, which is non-protein N (NPN), was calculated as CP minus TP precipitated with tungstic acid. Fraction B1 is TP soluble in borate-phosphate buffer. Fraction B2 is buffer insoluble CP minus CP insoluble in neutral detergent. Fraction B3 is NDICP minus ADICP, the latter also representing fraction C.

As neutral detergent fibre (NDF) values of the feed samples that were determined within the CP fractionation schedule by manual filtration on paper according to the recommendations of Licitra et al. (1996) may deviate from those obtained with the conventional NDF method, the cell wall fraction obtained as a residue on filter paper was named PNDF.

Values for CP, CP fractions, and PNDF were used to estimate RUP for assumed ruminal passage rates (K_p) of 0.02, 0.05, and 0.08/h (designated RUP2, RUP5, and RUP8, respectively; in g/kg CP) according to regression equations of Shannak et al. (2000).

2.4 Intestinal digestibility of ruminally undegraded crude protein

Intestinal digestibility of RUP was estimated via a three-step enzymatic *in vitro* method [Hippenstiel et al., 2015]. Basically, the method corresponds to Calsamiglia and Stern (1995), except that ruminal incubation was replaced by application of *Streptomyces griseus* protease [Licitra et al., 1998). True protein content was determined according to Licitra et al. (1996) using trichloroacetic acid (TCA) as a precipitating agent. To simulate rumen incubation, samples (2.5 g) were accurately weighed into 500-ml Erlenmeyer flasks and 200 ml of borate-phosphate buffer (pH 6.7–6.8) were added. Flasks were then kept in a shaking water bath at 39°C for 1 h. Protease solution consisting of the same borate-phosphate buffer and *S. griseus* protease powder (1.0 U/ml; Type XIV, ≥ 3.5 U/mg solid, P5147, Sigma-Aldrich, St. Louis, MO, USA) was added in an amount corresponding to 41 U/g TP [Licitra et al. 1998). After 18 h of incubation as recommended by Licitra et al. (1998) for concentrates, contents were filtered through a fibre bag (30 µm pore size, Gerhardt, Königswinter, Germany). Fibre bags were washed in a beaker with fresh deionized water ten times before residuals were freeze dried and analysed for N content using the Dumas combustion method (method 4.1.2, VDLUFA 2012; FP328, Leco 8.1, Leco, Mönchengladbach, Germany).

Residuals were weighed into 50 ml centrifugation tubes in an amount including 15 mg N. After addition of 10 ml of a 0.1 N HCl solution (pH 1.9) containing 1 g/l of pepsin (P7012, Sigma-Aldrich, St. Louis, MO, USA), tubes were incubated for 1 h in a shaking water bath at 38°C. Subsequently, the solution was neutralized with 0.5 ml of 1 N NaOH. Then, 13.5 ml of a phosphate buffer (pH 7.8) containing 37.5 mg of pancreatin (P7545, Sigma-Aldrich, St. Louis, MO, USA) were added to each tube. After incubation for 24 h and vortexing every 8 h, 3 ml of TCA (1000 g/l) were pipetted to each tube to stop or minimise enzymatic action and precipitate undigested protein. After about 15 min, the samples were centrifuged at 10,000 *g* for 15 min at 5°C. The precipitate was then filtered through filter paper (388, Munktell, Falun, Sweden), and the residue on the filter paper was analysed for insoluble N by the Kjeldahl procedure (method 4.1.1, VDLUFA, 2012). Protease and pepsin-pancreatin incubations were carried out in eight replications. Intestinal protein digestibility was calculated as N soluble in TCA divided by 15 mg N.

2.5 Data treatment and statistical analysis

A face-centered central composite design (CCD) consisting of a 2^3 full factorial design, four center points and six axial points ($\alpha = 1$) was applied to investigate the effects of three independent variables (pH, reaction time and drying temperature) on the dependent variables (CP fractions A, B1, B2, B3 and C, and the resulting RUP2, RUP5 and RUP8 values). This design resulted in 18 alkali-treated samples including four replicates of a randomly selected single sample providing the four center points (Table 2.2). All samples were prepared following a randomized run order. Design Expert Version 9.0.3.1 (State-Ease Inc., Minneapolis, MN, USA) was used for data analysis.

The data obtained experimentally and mathematically were fitted with a second order polynomial equation:

$$Y = \beta_0 + \sum_{i=1}^{k} \beta_i X_i \sum_{i=1}^{k} \beta_{ii} X_i^2 + \sum_i \sum_{j=i+1} \beta_{ij} X_i X_j,$$

where Y is the predicted response, β_0, β_i, β_{ii}, β_{ij} represent the model coefficients for intercept, linear, quadratic and interaction terms, X_i and X_j are the independent variables. K is the number of factors. The adequacy of the models was determined by evaluating the lack of fit and the coefficients of regression (R^2) obtained from the analysis of variance (ANOVA). Statistical significance of the model and model variables was determined at the 5% probability level ($p<0.05$). Based on the established regression models, a multi-response optimization was conducted with RUP2, RUP5 and RUP8 values set to maximum, while the pH values were set to target pH 7, 8.5 and 10, respectively. The CP fractions A, B1, B2, B3, and C could vary in the given ranges without specified targets.

3 Results

3.1 Chemical composition

The untreated SEM contained 926 g/kg dry matter and on dry matter basis 78.1 g/kg ash, 399 g/kg crude protein, 20.4 g/kg crude lipids, 344 g/kg aNDFom, 258 g/kg ADFom and 63.4 g/kg ADL.

3.2 Color development

The treatment of SEM resulted in green discoloration of all suspensions, showing absorption maxima at 685 nm. At pH 7, only a slight discoloration was observed, whereas the green color intensified with increasing pH as shown in Figure 2.1.

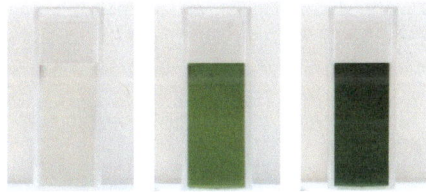

Figure 2.1: Development of green discoloration of solvent-extracted sunflower meal at (left to right) pH 7, pH 8.5 and pH 10. Sample colors intensified with increasing pH

3.3 Statistical relationships

Equations and determination coefficients of the response variables, namely the CP fractions A, B1, B2, B3 and C as well as the resulting RUP values (RUP2, RUP5 and RUP8), are given in Table 2.1. Homoscedasticity and normal distribution were assessed and given for all data so that no transformation was needed. The resulting response surface plots are given in Fig. 2.2.

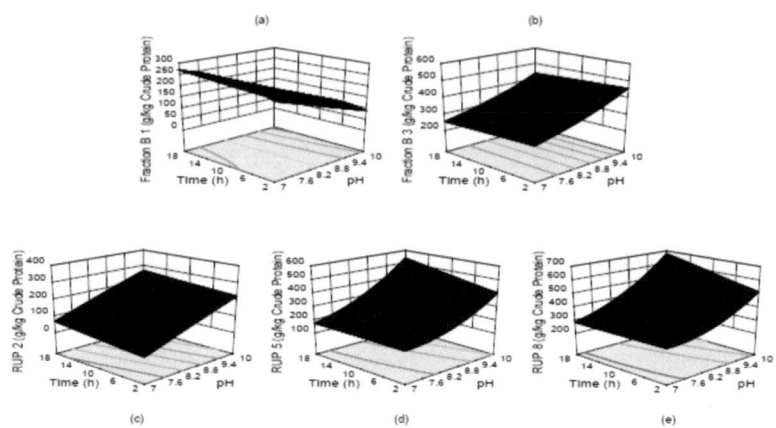

Figure 2.2: Significant influences by reaction parameters on single crude protein fractions. Increasing pH resulted in (a) decreasing B1 values, (b) increasing B3 values and, (c–e) higher ruminally undegraded crude protein (RUP) values

Table 2.1: Equations in actual factors and determination coefficients of the response variables. X_1: pH; X_2: reaction time

Response variable	Equation	R^2	R_{adj}	R_{pred}
A	$Y = 125.51 - 11.34X_1 - 0.07X_2 + 0.11X_1X_2$	0.59	0.50	0.25
B1	$Y = -420.17 + 19093X_1 + 14.86X_2 - 1.84X_1X_2 - 13.91X_1^2$	0.99	0.94	0.87
B2	$Y = 173.26 + 2.01X_1$	0.35	0.31	0.13
B3	$Y = 0.07873 - 0.00492X_1 + 0.00022X_2$	0.79	0.79	0.78
C	$Y = 0.00000998 - 0.00000068X_1$	0.35	0.31	0.35
RUP2	$Y = 449.62 - 66.53X_1 - 2.96X_2 + 0.07X_1X_2 + 3.04X_1^2$	0.87	0.86	0.77
RUP5	$Y = 1324.69 - 329.24X_1 - 19.39X_2 + 2.79X_1X_2 + 23.54X_1^2$	0.94	0.93	0.92
RUP8	$Y = 1599.01 - 363.54X_1 - 25.33X_2 + 3.55X_1X_2 + 25.31X_1^2$	0.96	0.94	0.91

3.4 Crude protein fractions

For fractions B1 and B3, the corresponding probability values indicate that the models were significant (p<0.0001) and had coefficients of determination (R^2) of 0.78 and 0.95, respectively.

Fraction B1 (Fig. 2 a) was influenced by reaction time and pH ($p<0.0001$). The pH had a quadratic effect, with increasing pH resulting in decreasing B1 values. Furthermore, the pH interacted with reaction time ($p<0.0001$) such that the positive linear effect of reaction time on fraction B1 was not observed at pH 7 but intensified with increasing pH.

Fraction B3 (Fig. 2 b) was influenced by pH and drying temperature ($p<0.0001$) with a positive linear effect of pH, whereas increasing drying temperatures resulted in decreasing B3 values. For fractions A, B2 and C, the regression coefficients were poor (0.35–0.59). Fraction C values ranged between 60 and 110 g/kg CP, with increasing pH values resulting in higher percentages of fraction C.

3.5 Ruminally undegraded crude protein

A wide range of RUP values was observed, ranging from 31 g/kg of CP for untreated SEM at an assumed rumen solid outflow rate of 0.02/h (RUP2) to 694 g/kg of CP for the alkaline treated SEM (pH 10, 18 h, 70°C) at a passage rate of 0.08/h (RUP8). The individual values are given in Table 2.2.

The alkaline treatment of SEM resulted in higher RUP values in all samples, hereby showing a positive correlation between increasing pH and RUP (Fig. 2 c-e). These findings were confirmed by significant regression models for RUP2, RUP5, and RUP8, respectively ($p<0.0001$, $R^2=0.87–0.96$), reflecting influences of pH (quadratic response) and reaction time (positive linear relationship). Furthermore, an interaction between pH and reaction time was observed: at pH 7, the reaction time had no influence on the estimated RUP values, whereas at pH 10 an increase in reaction time resulted in increasing RUP values.

Table 2.2: Values for ruminally undegraded crude protein (RUP) for assumed ruminal passage rates of 0.02, 0.05, and 0.08/h (RUP2, RUP5, RUP8; in g/kg CP) of sunflower meal samples before (untreated) and after alkali treatment at different pH, for varying reaction times and at different drying temperatures. Reaction parameters are sorted according to their impact on the resulting RUP values. The combination of pH 8.5, reaction time 10 h and drying temperature 85 °C provided the center points of the experimental central composite design and is therefore included four times.

pH	Reaction [h]	Drying [°C]	RUP2	RUP5	RUP8
6.8	–	–	31	125	233
7	2	100	39	123	255
7	2	70	53	148	252
7	10	85	61	172	282
7	18	100	61	184	303
7	18	70	88	226	344
8.5	2	85	123	249	357
8.5	10	85	126	259	364
8.5	10	85	126	264	376
8.5	10	70	113	266	393
8.5	10	85	121	269	388
8.5	10	85	125	271	391
8.5	18	85	129	278	396
8.5	10	100	141	280	386
10	2	70	185	365	477
10	2	100	246	432	543
10	10	85	249	461	584
10	18	70	263	528	694
10	18	100	362	574	685

3.6   Optimization and verification of predictive models

Three of alkali-treated additional samples SEM were prepared according to the results of the multi-response optimization, which led to an optimal reaction time of 18 h for alkaline treatment in each case, resulting from the positive linear relation between time and RUP values. Since the drying temperature of the treated SEM did not show any effects on the resulting RUP values but affected fraction B3, all three additional samples were dried at 70 °C to avoid any heat-induced damage of SEM protein. As exemplified in Fig. 2.3, SEM treatment led to a shift of the CP fractions from B1 to B3 with increasing pH and a concomitant increase in RUP values. However, increasing pH from 8.5 to 10 may not further elevate RUP values.

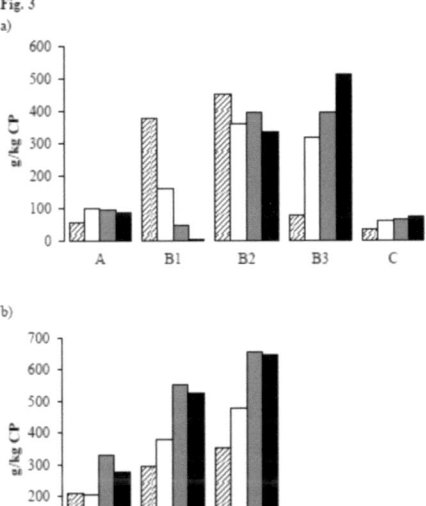

Figure 2.3: Characteristics of sunflower meal either untreated (▨) or treated according to results of multi-response optimization, i.e. undergoing 18 h of alkali treatment at pH 7 (□), 8.5 (▣), and 10 (■) and subsequent drying at 70 °C; a) Crude protein fractions according to Licitra et al. (1996). b) Values for ruminally undegraded crude protein (RUP) for assumed ruminal passage rates of 0.02, 0.05, and 0.08/h (RUP2, RUP5, RUP8; estimated according to Shannak et al., 2000)

3.7 Estimation of intestinal digestibility

Estimated intestinal digestibilities including laboratory standard deviations of the three samples treated according to optimized parameters are shown in Table 2.3. Results indicate an increase in the intestinal digestibility of the alkali-treated samples compared to untreated SEM.

Table 2.3 Intestinal digestibility (± laboratory standard deviation) of ruminally undegraded crude protein (RUP) in four samples of sunflower meal either untreated or treated according to results of multi-response optimization, i.e. undergoing 18 h of alkali treatment at different pH values and subsequent drying at 70°C

pH	Intestinal digestibility (g/kg of RUP)
Untreated[†]	666 ± 43
pH 7	740 ± 20
pH 8.5	731 ± 16
pH 10	746 ± 30

[†]Untreated, no addition of NaOH or HCl.

4 Discussion

In the present study, SEM was subjected to alkaline treatment to investigate the impact of interactions between SEM protein and CQA on ruminal CP degradation and, for selected samples, intestinal digestibility of RUP. By alkalization, the oxidation of CQA and thus covalent binding of CQA quinones to the nucleophilic side chains of proteins with green discoloration as a result can be expected (Prigent, 2005; Le Bourvellec and Renard, 2012; Yabuta et al., 2001).

Indeed, alkalization of SEM resulted in green color of all treated solutions, intensifying with increasing pH. A slight discoloration occurred even at pH 7, indicating that pH dependent protein CQA interactions took place not only in alkalized solutions but, to some extent, at neutral reaction conditions. This effect was not observed in suspensions without any addition of NaOH (pH 6.8). Thus, it can be assumed that the addition even of minor amounts of NaOH resulted in the oxidation of CQA to its *o*-quinone and the subsequent formation of green reaction products, which may constitute of benzacridin derivatives according to our recent investigations [Bongartz et al., 2016].

The CNCPS fractionation schedule showed that the alkaline treatment of SEM led to a shift of the CP fractions from the fast degradable B1 fraction to the slowly degradable B3 fraction. These findings are in agreement with previous studies showing a similar shift of the CP fractions from B1 to B3 in red clover (*Trifolium pratense* L.) rich in *o*-quinones [Grabber and Coblentz, 2009].

Related shifts have been observed in rapeseed meal which had been chemically treated with formaldehyde as well as soybeans that had been subjected to superheated steam [Shannak et al., 2000].

Based on investigations into heat-processed and unprocessed plant proteins, namely sunflower, cottonseed and soybean cakes as well as whole soybean, whole sunflower seed, whole cottonseed and whole lupin seed, Schroeder et al. (1996) concluded that fraction C values exceeding a limit of 120 to 150 g/kg CP indicate protein damage due to the formation of Maillard products under thermal treatment. Such heat damages can result in a significantly lower intestinal digestibility and subsequent amino acid absorption and availability, thereby reducing the quality of the original feed. In all samples, fraction C had a share of 60 to 110 g/kg CP. Therefore, significant protein damage and, consequently, any related quality reduction of the feed protein, can be ruled out [Licitra et al. 1996; Schroeder et al., 1996].

Because the drying temperature in our study did not exceed 100°C, our results are also in line with Dakowski et al. (1996), who reported that processing of rapeseed meal only resulted in overprotection from ruminal CP degradation and lowered intestinal CP digestibility when the processing temperature exceeded 130°C.

Furthermore, alkalization of SEM led to a considerable increase in RUP in all samples, with factors up to 11.7 for RUP2, 4.6 for RUP5 and 3.0 for RUP8, indicating that the impact of alkaline treatment on RUP is highest at low passage rates. A slight drop in the RUP2 values of the optimized sample at pH 7 compared to the RUP2 value at pH 8.5 was not confirmed by the results of the CCD analyses, suggesting that this value was a single outlier.

To evaluate the intestinal digestibility, samples were subjected to a three step enzymatic *in vitro* procedure following Hippenstiel et al. (2015).

The intestinal digestibility of the SEM was enhanced by alkalization with an increase by approximately 10%, in comparison to untreated SEM. Although, based on the considerations reported above, we did not expect a reduction in intestinal CP digestibility in response to alkaline treatments at different intensities (pH, temperature, time), the observed increase in intestinal CP digestibility with increasing processing regime was unexpected and is perhaps the most striking observation of this study. Typically, a near-to-optimum situation is achieved if an increase in RUP values does not result in lowered values for intestinal CP digestibility (expressed as fraction of RUP entering the small intestine), such that protein (amino acid) supply to the animal raises proportionally to RUP increment. Often, an increasing intensity or strength of processing conditions, although increasing RUP values, lowers intestinal CP digestibility. A case in point is Can et al. (2011), who studied treatment combinations (temperature, duration and xylose supplementation prior to treatment) on *in vitro* ruminal degradability and intestinal protein and total CP digestibility of untreated and differently treated soybean (SBM) and cottonseed (CSM) meals. The control SBM had higher *in vitro* ruminal CP degradability values than the treated samples. Intestinal protein digestibility and total-tract CP digestibility of CSM and SBM were affected by the treatment. The results of the study indicated that not only ruminal CP degradability is reduced but also intestinal and total tract CP digestibilities may be lowered depending on protein source and intensity of the processing conditions.

If, however, an increase in RUP (expressed as proportion of feed CP) is paralleled by an increase in intestinal CP digestibility (expressed as fraction of RUP) as observed in this study, the gain in protein that can be digested in the small intestine would be disproportionately high and, therefore, would provide potential for elevating RUP values without a discount on post-ruminal digestibility. Further studies should cover this issue in more detail on a larger set of feed materials.

5   Conclusion

The induced formation of CQA quinones by alkaline treatment led to a reaction with SEM protein, which entailed a considerably reduced simulated degradation of the protein by ruminal microorganisms. This resulted in enhanced RUP values with factors ranging from approximately 3 (RUP8) to 12 (RUP2). Furthermore, the intestinal digestibility of the alkali-treated SEM was enhanced by approximately 10%. In summary, a directed alkalization of SEM may indeed be an interesting approach to modify feed CP in a way that highly digestible RUP is provided to directly supply ruminants with amino acids in the small intestine. This will be particularly relevant if rumen microbial synthesis is limited by energy supply, e.g. at the onset of lactation. Further research will be dedicated to the structure elucidation of the reaction products and implications of these findings in food and feed quality and safety.

References

Antoniewicz, A.; van Vuuren, A.; van der Koelen, C.; Kosmala, I. 1992: Intestinal digestibility of rumen undegraded protein of formaldehyde-treated feedstuffs measured by mobile bag and in vitro technique. Anim. Feed Sci. Technol. 39, 111–124.

Bongartz, V.; Brandt, L.; Gehrmann, M.L.; Zimmermann, B.F.; Schulze-Kaysers, N.; Schieber, A. 2016: Evidence for the formation of benzacridine derivatives in alkaline-treated sunflower meal and model solutions. Molecules 21, 91 (10.3390/molecules21010091).

Calsamiglia, S.; Ferret, A.; Reynolds, C.; Kristensen, N.; van Vuuren, A. 2010: Strategies for optimizing nitrogen use by ruminants. Animal 4, 1184–1196.

Calsamiglia, S.; Stern, M.D. 1995: A three-step in vitro procedure for estimating intestinal digestion of protein in ruminants. J. Anim. Sci. 73, 1459–1465.

Can, A.; Hummel, J.; Denek, N.; Südekum, K.-H. 2011: Effects of non-enzymatic browning reaction intensity on in vitro ruminal protein degradation and intestinal protein digestion of soybean and cottonseed meals. Anim. Feed Sci. Technol. 163, 255–259.

Dakowski, P.; Weisbjerg, M.R.; Hvelplund, T. 1996: The effect of temperature during processing of rape seed meal on amino acid degradation in the rumen and digestion in the intestine. Anim. Feed Sci. Technol. 58, 213–226.

González-Pérez, S.; Merck, K.; Vereijken, J. 2002: Isolation and characterization of undenatured chlorogenic acid free sunflower (*Helianthus annuus*) proteins. J. Agric. Food Chem. 50, 1713–1719.

Grabber, J.; Coblentz, W. 2009: Polyphenol, conditioning, and conservation effects on protein fractions and degradability in forage legumes. Crop Sci. 49, 1511–1522.

Hippenstiel, F.; Kivitz, A.; Benninghoff, J.; Südekum, K.-H. 2015: Estimation of intestinal protein digestibility of protein supplements for ruminants using a three-step enzymatic in vitro procedure. Arch. Anim. Nutr. 69, 310–318.

Le Bourvellec, C.; Renard, C. 2012: Interactions between polyphenols and macromolecules: quantification methods and mechanisms. Crit. Rev. Food Sci. Nutr. 52, 213–248.

Licitra, G.; Hernandez, T.M.; Van Soest, P.J. 1996: Standardization of procedures for nitrogen fractionation of ruminant feeds. Anim. Feed Sci. Technol. 57, 347–358.

Licitra, G.; Lauria, F.; Carpino, S.; Schadt, I.; Sniffen, C.J.; Van Soest, P.J. 1998: Improvement of the *Streptomyces griseus* method for degradable protein in ruminant feeds. Anim. Feed Sci. Technol. 72, 1–10.

Lomascolo, A.; Uzan-Boukhris, E.; Sigoillot, J.-C.; Fine, F. 2012: Rapeseed and sunflower meal: a review on biotechnology status and challenges. Appl. Microbiol. Biotechnol. 95, 1105–1114.

Patra, A.; Saxena, J. 2009: Dietary phytochemicals as rumen modifiers: a review of the effects on microbial populations. Anton. Leeuw. Int. 96, 363–375.

Prigent, S. 2005: Interactions of phenolic compounds with globular proteins and their effects on food-related functional properties. Ph.D. Thesis, Wageningen University, The Netherlands.

Rawel, H.; Meidtner, K.; Kroll, J. 2005: Binding of selected phenolic compounds to proteins. J. Agric. Food Chem. 53, 4228–4235.

Schroeder, G.; Erasmus, L.; Leeuw, K.-J.; Meissner, H. 1996: The use of acid detergent insoluble nitrogen to predict digestibility of rumen undegradable protein of heat processed plant proteins. S. Afr. J. Anim. Sci. 26, 49–52.

Shannak, S.; Südekum, K.-H.; Susenbeth, A. 2000: Estimating ruminal crude protein degradation with *in situ* and chemical fractionation procedures. Anim. Feed Sci. Technol. 85, 195–214.

Sniffen, C.J.; O'Connor, J.D.; Van Soest, P.J.; Fox, D.G.; Russell, J.B. 1992: A net carbohydrate and protein system for evaluating cattle diets: II. Carbohydrate and protein availability. J. Anim. Sci. 70, 3562–3577.

VDLUFA, 2012. Verband Deutscher Landwirtschaftlicher Untersuchungs- und Forschungsanstalten. Handbuch der Landwirtschaftlichen Versuchs- und Untersuchungsmethodik (VDLUFA-Methodenbuch), Bd. III. Die Chemische Untersuchung von Futtermitteln. VDLUFA-Verlag, Darmstadt, Germany.

Weisz, G.; Schneider, L.; Schweiggert, U.; Kammerer, D.; Carle, R. 2010: Sustainable sunflower processing – I. Development of a process for the adsorptive decolorization of sunflower (*Helianthus annuus* L.) protein extracts. Innov. Food Sci. Emerg. Technol. 11, 733–741.

Wu, S.; Papas, A. 1997: Rumen-stable delivery systems. Adv. Drug Delivery. Rev. 28, 323–334.

Yabuta, G.; Koizumi, Y.; Namiki, K.; Hida, M.; Namiki, M. 2001: Structure of green pigment formed by the reaction of caffeic acid esters (or chlorogenic acid) with a primary amino compound. Biosci. Biotechnol. Biochem. 65, 2121–2130.

Yu, P.; Goelema, J.; Leury, B.; Tamminga, S.; Egan, A. 2002: Using the DVE/OEB model to determine optimal conditions of pressure toasting on horse beans (*Vicia faba*) for the dairy feed industry. Anim. Feed Sci. Technol. 99, 141–176.

Chapter 3: Evidence for the Formation of Benzacridine Derivatives in Alkaline-Treated Sunflower Meal and Model Solutions[*]

Sunflower extraction meal (SEM) is an economically interesting protein source. During alkaline extraction of proteins, the presence of chlorogenic acid (CQA) in the meal gives rise to the formation of o-quinones. Reactions with nucleophiles present in proteins can lead to green discoloration. Although such reactions have been known for a long time, there is a lack of information on the chemical nature of the reaction products. SEM and model systems consisting of amino acids and CQA were subjected to alkaline treatment and, for comparison, to oxidation of CQA by polyphenoloxidase (PPO). Several green trihydroxy benzacridine (TBA) derivatives were tentatively identified in all samples by UHPLC-DAD-MS/MS. Surprisingly, in alkaline-treated samples of particular amino acids as well as in SEM, the same six TBA isomers were detected. In contrast, the enzymatically oxidized samples resulted in only three TBA derivatives. Contrary to previous findings, neither peptide nor amino acid residues were attached to the resultant benzacridine core. The results indicate that the formation of TBA derivatives is caused by the reaction between CQA quinones and free NH2 groups. Further research is necessary to elucidate the structure of the addition products for a comprehensive evaluation of food and feed safety aspects.

Keywords: amino acid; benzacridine derivatives; chlorogenic acid; green color; oxidation; o-quinone

---

[*] This chapter has been published:
Bongartz, V.; Brandt, L.; Gehrmann, M.L.; Zimmermann, B.F.; Schulze-Kaysers, N.; Schieber, A (2016): Evidence for the formation of benzacridine derivatives in alkaline-treated sunflower meal and model solutions. Molecules 21 (1), 91.

1 Introduction

Sunflower (*Helianthus annuus* L) seeds are considered a promising source of proteins. Except for their low lysine content, sunflower proteins match the FAO (Food and Agriculture Organization) reference protein patterns in terms of amino acid composition and are low in antinutritive compounds. In addition to their relatively high nutritive value, sunflower proteins display various interesting functional characteristics comparable to those of soybean and other legume proteins, such as emulsifying or foaming properties [Weisz et al., 2010].

The press residue originating after extraction of sunflower oils, in the following referred to as sunflower extraction meal (SEM), contains 40%–50% protein and is therefore an economically interesting source of nutrients [González-Pérez et al., 2002].

SEM is also rich in phenolic compounds, with chlorogenic acid (CQA) being the predominant component. Depending on environmental and genetic factors, total phenolics may range from 1%–4% [Weisz et al., 2010].

CQA has been associated with various health-promoting effects such as antioxidative, antibiotic or anti-inflammatory properties, which possibly prevent diseases associated with oxidative stress [Kammerer at al., 2007].

While highest protein yields are usually obtained by alkaline extraction, the presence of CQA in the meal gives rise to the formation of *o*-quinones under alkaline conditions. Additionally, *o*-quinones are formed in the presence of polyphenoloxidase (PPO) due to enzymatic oxidation. O-quinones are electron-deficient molecules that readily react with nucleophiles, such as thiols or amino groups present in proteins. In intact plants, proteins and CQA are located in separate compartments, whereas in mechanically processed plants, CQA and proteins can interact with each other due to the ruptured cell structure. Such interactions have been shown to alter the physicochemical properties of proteins [Pierpoint, 1969; Rawel et al., 2005, Le Bourvellec & Renard, 2012].

Furthermore, discoloration of SEM protein may occur, which limits its application in foods [González-Pérez et al., 2002].

Although green discoloration as a result of the reaction of quinones and proteins has long been realized, only little information is available on the chemical nature of the reaction products. In model solutions containing caffeic acid esters, Namiki et al. detected green benzacridine derivatives, formed after reaction of quinones and lysine [Namiki et al., 2001].

In the present study, SEM and model systems were subjected to alkaline treatment to investigate reactions between SEM protein and chlorogenic acid quinone. UHPLC-DAD-MS/MS analysis was used for the characterization of the resulting green adducts.

2 Results and Discussion

In the present study, SEM was subjected to alkaline treatment to investigate reactions between SEM protein and CQA quinone. Additionally, the 20 proteinogenic amino acids as well as NH_3 were tested for their reactivity towards alkaline and enzymatically generated CQA quinone and to characterize the resulting adducts by LC/MS. It was expected that such a comparatively simple model system would allow conclusions to be drawn as to the situation in the intact protein.

2.1 Color Development in SEM Extract and Model Solutions

As shown in Figure 3.1, the non-alkalized SEM extract remained colorless (a), whereas alkalization of SEM extract resulted in green discoloration (b). Corresponding discolorations of sunflower meal and other plant materials such as sweet potato (Ipomoea batatas) or greater burdock (Arctium lappa) in alkaline surroundings are known to occur due to CQA quinone interactions with proteins [Pedrosa et al., 2000; Yabuta et al. 2001].

In proteins, the ε-amino group of lysine and the thiol group of cysteine have been reported to be the preferred sites of reaction with quinones. Discoloration due to the addition of oxidized phenolic compounds to single amino acids such as tryptophan, histidine, and tyrosine has also been observed [Kroll et al., 2003; Prigent, 2005; Prigent et al., 2008].

Namiki et al. and Yabuta et al. reported that, upon oxidation, CQA quinone gives rise to the formation of a dimer, which subsequently reacts with the amino compound to finally yield a green benzacridine derivative [Namiki et al., 2001, Yabuta et al. 2001].

For the formation of such derivatives, Yabuta et al. concluded that an ortho-diphenol structure with a carbonyl side chain as present in CQA quinone is mandatory [Yabuta et al. 2001].

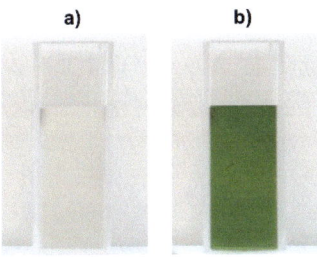

Figure 3.1: Extract of (a) non-alkalized SEM and (b) alkalized SEM (pH 9)

The incubation of CQA with the respective amino acid at pH 9 resulted in color formation including different shades of green, brown and, in the case of tryptophan, red (Figure 3.2). All peaks showed absorption maxima at 631 and 461 nm. Thus, it can be assumed that the corresponding compounds are responsible for the coloration of the solutions. Since a green pigment had been observed upon reaction of primary amino groups with a dimer of oxidized ethyl caffeate [Namiki et al., 2001, Yabuta et al. 2001], the green color may indicate attachment of CQA dimers to the α-NH_2 group in all model systems, except for proline and cystein. Due to its highly nucleophilic character, cysteine reacted with CQA monomers instead of dimers and, consequently, did not form any colored adducts [Prigent, 2005; Prigent et al., 2008].

Among the tested amino acids, proline was the only amino acid with a secondary amino group. The reaction mixture with proline and CQA showed the same shade of brown as the CQA-solution without any added amino compounds. As indicated by Yabuta, a primary NH_2 group is necessary for a reaction between a CQA-dimer and an amino acid [Yabuta et al. 2001].

Therefore, proline could not react with CQA to trihydroxy benzacridine (TBA). The red color in the alkaline reaction mixture with tryptophan and CQA can be explained by the presence of the indole group of tryptophan. Contrary to alkalization, oxidation by PPO at pH 7 resulted in dark brown discoloration in all samples that resembled the brown color of oxidized CQA solution (see Figure 3.2). CQA quinones are known to readily polymerize, which leads to the formation of melanines. As incubation with PPO results in a large amount of o-quinones, it is reasonable to assume that the dark color is caused by melanines [Namiki et al., 2001, García Cárnovas et al., 1982; Clifford et al., 2003].

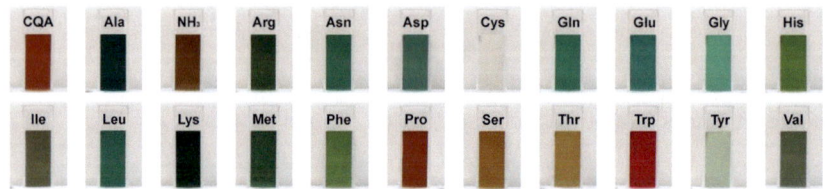

Figure 3.2: Color development in model systems containing CQA and individual amino acids at pH 9

2.2 UHPLC-DAD-MS/MS Analysis

UHPLC-DAD-MS/MS analysis was used for the characterization of the resulting colored adducts. SIM (selected ion monitoring) was applied to detect specific molecule ions as derived from Prigent et al. and Schilling et al.

Positive ionization yielded higher intensities and was therefore used for further fragmentation experiments. Although only 5-caffeoylquinic acid was used initially, incubation at 40 °C led to the formation of 3- and 4-caffeoylquinic acids that were identified by their characteristic fragmentation pattern as described by Clifford et al. [Clifford et al., 2003].

With respect to the objectives of the present study, it was of greater interest that in all incubation mixtures a large number of peaks of low intensity were detected that corresponded to CQA dimers as described by Schilling et al.

The CQA dimers are known to be the main compounds reacting with amino acids, as also observed for caffeic acid esters by Namiki and co-workers. It should be noted that several structures have been postulated for caffeic acid and CQA dimers, but only a few have been structurally elucidated [Namiki et al., 2001, Yabuta et al. 2001,15].

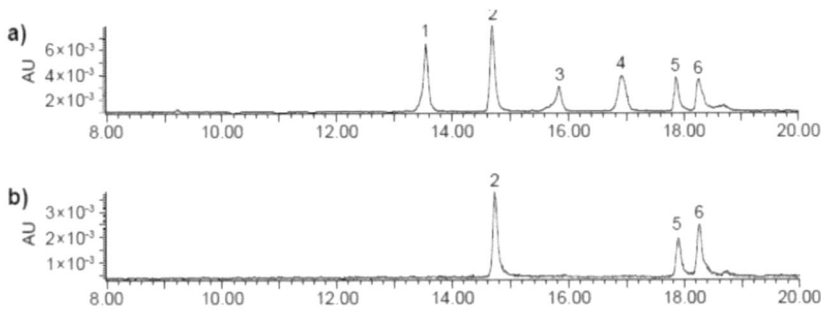

Figure 3.3: Exemplary UHPLC-DAD chromatograms of TBA isomers (450 nm) in (a) a model system containing α-alanine and CQA at pH 9 and in (b) a model solution containing α-alanine and PPO at pH 7

Although several distinct peaks were detected, in alkalized SEM extract and all alkaline model systems except for cysteine and proline, six peaks were dominant. Corresponding reaction mixtures containing PPO at pH 7 yielded three peaks that equaled three of the six peaks in the alkaline solutions (Figure 3.3). The mass spectrometric characteristics of the respective peaks are shown in Tables 3.1 and 3.2. The latter findings are in agreement with Schilling et al. who detected three equivalent peaks in CQA solution containing N-BOC-lysine (N-butoxycarbonyl-lysine) and PPO. PPO is more specific and leads to fewer oxidation products than alkaline conditions. In alkaline solutions, 5-CQA isomerizes to 3-CQA and 4-CQA. Thus, the formation of CQA isomers is accounted for the occurrence of six peaks in alkaline solutions as opposed to three peaks in solutions containing PPO. All peaks displayed an *m/z* ratio of 700, matching fragments of corresponding reaction adducts as described by Schilling et al. The respective fragmentation pathway is given in Figure 3.4.

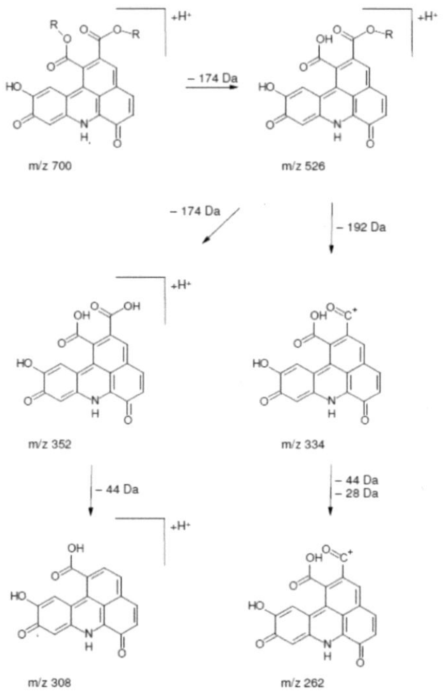

Figure 3.4: Postulated fragmentation pathway of a chlorogenic L-lysine adduct. R = Quinic acid [Modified from Schilling et al., 2008]

Table 3.1: Mass spectrometric characteristics of trihydroxy benzacridine (TBA) isomers found in alkalized SEM extract. Model systems containing chlorogenic acid (CQA) and individual L-amino acids or NH_3 at pH 9 showed identical peaks in terms of retention time and m/z. Corresponding model solution at pH 7 containing polyphenoloxidase (PPO) displayed only peaks 2, 5 and 6

Peak	R_t (min)	[M+H]	UHPLC/ESI(+)-MS^2 fragments
1	13.51	700.3	MS^2[700.3]: 334 (100), 352 (54), 526 (11), 262 (7), 308 (7), 654 (5) 700 (3)
2	14.65	700.3	MS^2[700.3]: 334 (100), 352 (83), 526 (18), 262 (5), 308 (9), 654 (4), 700 (6)
3	15.80	700.3	MS^2[700.3]: 334 (100), 352 (50), 526 (16), 262 (6, 308 (4), 654 (1), 700 (4)
4	16.87	700.3	MS^2[700.3]: 334 (100), 352 (47), 526 (9), 262 (8), 308 (7), 654 (1), 700 (5)
5	17.79	700.3	MS^2[700.3]: 334 (100), 352 (48), 526 (7), 262 (5), 308 (5), 654 (4) 700 (1)
6	18.20	700.3	MS^2[700.3]: 334 (100), 352 (60), 526 (14), 262 (9), 308 (6), 654 (3), 700 (2)

Table 3.2: Mass spectrometric characteristics of additional TBA isomers found in the alkalized model system containing CQA and L-lysine. All peaks showed absorption maxima at 631 and 461 nm

Peak	R_t (min)	[M+H]	UHPLC/ESI(+)-MS^2 fragments
1	13.48	829	MS^2[829]: 334 (100), 128 (83), 352 (20), 526 (29), 336 (27), 654 (7) 700 (5), 130 (19), 318 (4), 480 (6)
2	14.59	829	MS^2[829]: 334 (100), 128 (76), 352 (25), 526 (23), 336 (19), 654 (2) 700 (3), 130 (26) 318 (6), 480 (7)
3	15.73	829	MS^2[829]: 334 (100), 128 (96), 352 (27), 526 (40), 336 (52), 654 (3), 700 (22), 130 (9), 318 (11), 480 (5),
4	16.74	829	MS^2[829]: 334 (100), 128 (84), 352 (51), 526 (36), 336 (40), 654 (3), 700 (17), 130 (16), 318 (7), 480 (5)
5	17.73	829	MS^2[829]: 334 (100), 128 (80), 352 (48), 526 (30), 336 (18), 654 (8), 700 (21), 130 (14), 318 (10), 480 (3),
6	18.11	829	MS^2[829]: 334 (100), 128 (78), 352 (54), 526 (36), 336 (16), 654 (4), 700 (16), 130 (15), 318 (10), 480 (5),

Additionally, in the model solution containing lysine and PPO, three peaks with m/z 829 were observed that were dominant compared to the three peaks with m/z 700. Schilling et al. detected TBA derivatives with m/z 929 in a N-BOC-lysine solution, which fragmented to a molecule with m/z 700 due to the loss of the BOC-lysine residue (Δ 229). In this study, lysine, with a molecular mass of 146, was used instead of BOC-lysine, which also fragmented to a molecule with m/z 700 (Figure 3.5).

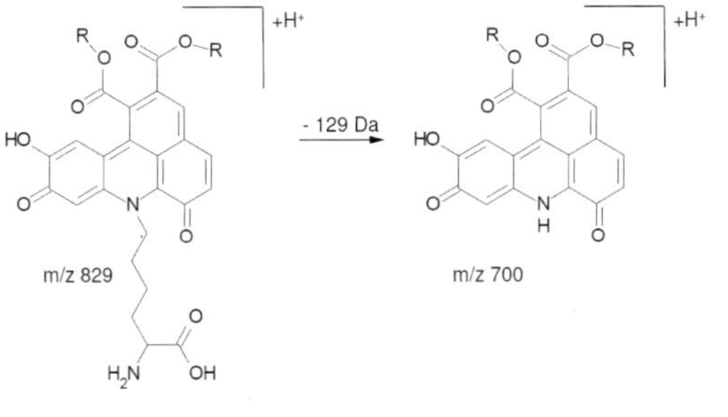

Figure 3.5: Postulated fragmentation of the chlorogenic L-lysine adduct with m/z 829 to the chlorogenic L-lysine adduct with m/z 700. R = Quinic acid. Modified from [Schilling et al., 2008]

Therefore, the fragment at m/z 829 detected in the model solution containing lysine and PPO is explained by a TBA containing a lysine residue in agreement with the TBA-BOC-lysine adduct described by Schilling et al. [Schilling et al., 2008].

The occurrence of peaks with m/z 700 in the same solution is explained by the additional NH_2 group of lysine: the presence, or lack, of the lysine residue in the TBA molecule may depend on which NH_2 group is incorporated in the TBA core. In all other model systems, exclusively peaks with m/z 700 were detected, i.e., TBA cores without any side chains attached. All compounds showed a similar fragmentation pattern (Tables 3.1 and 3.2), regardless if the $[M + H]^+$ ion had m/z 700 or m/z 829. Although all peaks with m/z 700 were nearly identical in terms of fragmentation, they were detected at different retention times, suggesting that the corresponding compounds were structural isomers. These findings contradict Namiki et al., who proposed a reaction mechanism for the formation of TBA derivatives with the corresponding amino acid side chain attached to the resulting benzacridine core [Namiki et al., 2001].

It should be mentioned that Namiki et al. proposed the reaction mechanism on the base of n-butylamine, whereas Schilling et al. used N-BOC-lysine, thereby blocking the α-NH_2 group of the amino acid. In this study, the α-NH_2 group of all tested amino acids was available for the reaction as proposed by Namiki et al. for n-butylamine. To verify that the reaction of the α-NH_2 group is accountable for the loss of the amino acid side chain, β-alanine was used in additional CQA model systems. Neither alkaline nor enzymatic treatment resulted in peaks with m/z 700. Instead, in the enzymatically treated β-alanine solution, three peaks with m/z 772 were detected, accounting for TBAs with an alanine side chain attached (Figure 3.6). Furthermore, in the alkaline treated β-alanine solution, six peaks with m/z 772 were detected, which corresponded to the three or six peaks, respectively, shown in the respective α-alanine systems.

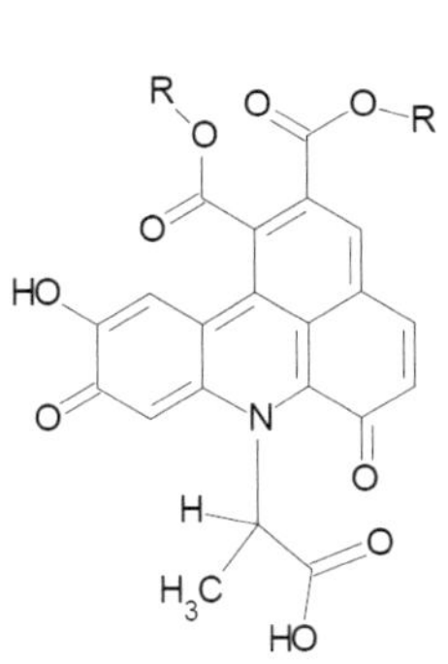

Figure 3.6: Postulated chlorogenic β-alanine adduct with *m/z* 772 in alkaline and enzymatically treated solutions. R = Quinic acid

These findings suggest that the presence, or lack, of an amino acid residue in the resulting TBA core depends on which NH_2 group is involved in the TBA formation. If β- or ε-NH_2 groups are incorporated in the TBA core, it can be assumed that the amino acid residue remains attached. This presumption is confirmed by recent research regarding milk and coffee proteins, in which ε-NH_2 groups of lysine were found to be incorporated in proteins that had been modified with phenolics [Ali et al., 20012; Ali et al., 2013].

Generalizing the findings of Namiki and Schilling, a corresponding reaction mechanism is proposed in Figure 3.7, which postulates TBA formation as a result of interaction between CQA and α-amino acids or amino acids with β- or ε-NH2 groups [Namiki et al., 2001, Schilling et al., 2008].

These findings suggest that the presence, or lack, of an amino acid residue in the resulting TBA core depends on which NH_2 group is involved in the TBA formation. If β- or ε-NH_2 groups are incorporated in the TBA core, it can be assumed that the amino acid residue remains attached. This presumption is confirmed by recent research regarding milk and coffee proteins, in which ε-NH_2 groups of lysine were found to be incorporated in proteins that had been modified with phenolics [Ali et al., 2012; Ali et al., 2013].

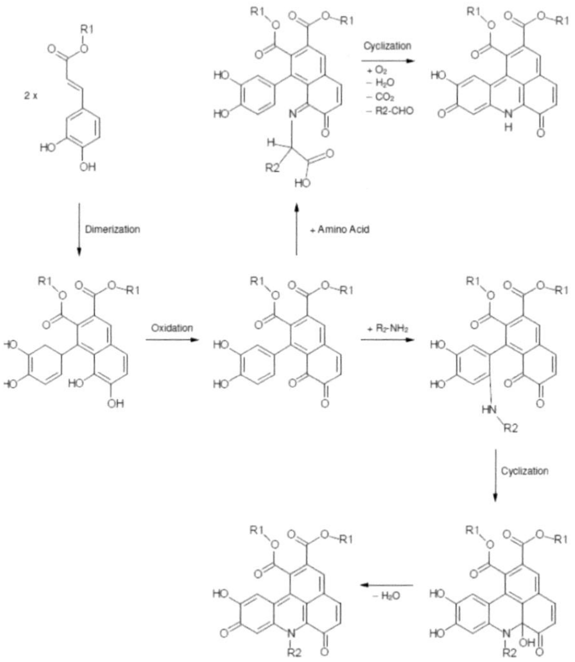

Chapter 3: Evidence for the Formation of Benzacridine Derivatives in Alkaline-Treated Sunflower Meal

Figure 3.7: Proposed reaction mechanism between CQA and individual amino acids. If an α-NH$_2$ group is involved, the resulting TBA core is free from any amino acid side chain. If TBA formation occurs at β- or ε-NH$_2$ groups, corresponding amino acid side chains remain attached. R1 = Quinic acid. R2 = amino side chain.

3 Experimental Section

3.1 Plant Material

Sunflower extraction meal pellets were purchased from Cefetra B.V. (Rotterdam, The Netherlands). The pellets were ground using a model RETSCH ZM centrifugal mill (Haan, Germany) at 200 s^{-1}.

3.2 Chemicals and Reagents

Solvents and reagents were purchased from VWR International (Darmstadt, Germany) and were of analytical, HPLC or MS grade. 5-O-Caffeoylquinic acid (5-CQA; >95%), amino acids (>98%) and mushroom tyrosinase (PPO) T3824 (1881 units/mg) were purchased from Roth (Karlsruhe, Germany), Rohm & Haas (Saint Priest, France) and Sigma-Aldrich (Steinheim, Germany), respectively. Deionized water was used throughout.

3.2.1 Sample Preparation– SEM Samples

3.2.1.1 Alkaline Treatment of SEM

Alkaline treatment of the ground SEM was carried out in a beaker at 5 s´1 using a Stuart orbital shaker (Staffordshire, UK) for 90 min. SEM and water were mixed at a meal-to-water ratio of 0.125 g¨ mL´1 . The suspension was adjusted to pH 9 with 1 M NaOH initially, and again after 10 and 30 min. To stop the reaction, the suspension was neutralized using 1 M HCl. The suspension obtained was dried for 12 h at 70 ˝C and ground again using the centrifugal mill at 200 s^{-1}.

3.2.1.2 Pressurized Liquid Extraction of SEM

Pressurized liquid extraction (PLE) was performed with a Dionex ASE 350 (Thermo Scientific, Idstein, Germany) system. For this purpose, 1 g SEM (before and after alkalization) were each mixed with diatomaceous earth (Thermo Scientific) and filled into 34 mL Dionex stainless steel cells. For extraction, the cell was filled with extraction solvent (water), pressurized (1500 psi), and heated (40 ˝C). Subsequently, the cell was rinsed with water (150% of the cell volume) and purged with a flow of nitrogen (100 psi, 60 s). Aliquots of 1 mL were filtered through 0.2 µm Chromafil RC-20/15 MS filters (Macherey-Nagel, Düren, Germany) and immediately stored at -80 °C until analysis.

3.2.1.3 Sample Preparation– Model Systems with 5-CQA and Amino Acids or Ammonia

For alkaline oxidized model systems, 112 mM aqueous solutions of each of the 20 α-amino acids (see Figure 3.1) or NH_3 or β-alanine and a 28 mM aqueous solution of CQA were mixed at a ratio of 1/1 (v/v) as reported by Prigent et al. [Prigent et al., 2008].

The solutions were adjusted to pH 9 with 0.1 M NaOH initially, and again after 10 and 30 min. The samples were stirred for 24 h at room temperature. 28 mM solutions of CQA mixed with water (1/1, v/v) served as a control. To stop the reaction, the solutions were neutralized using 1 M HCl. For enzymatic model systems, 28 mM aqueous solutions of each of the 20 amino acids or β-alanine or NH_3 containing PPO (15 U mL^{-1}) and a 112 mM aqueous solution of CQA were mixed at a ratio of 1/1 (v/v) [Prigent et al., 2008]. The solutions were adjusted to pH 7 with 0.1 M NaOH initially, and again after 10 and 30 min. The samples were stirred for 24 h at room temperature. To stop the reaction, the solutions were neutralized using 1 M HCl. Aliquots of 1 mL were filtered through 0.2 µm Chromafil RC-20/15 MS filters (Macherey-Nagel) and immediately stored at -80 °C until analysis.

3.3 UHPLC-DAD-MS/MS

An Acquity UPLC system from Waters (Milford, MA, USA) consisting of a binary pump (BSM), an autosampler (SM) cooled at 10 °C, a column oven (CM) set at 40 °C, a diode array detector (PDA) scanning from 190 to 500 nm, and an Acquity TQD triple-quadrupole mass spectrometer with an electrospray interface were used. The compounds were separated using an Acquity HSS-T3 RP18 column (150 mm ˆ 2.1 mm; 1.8 µm from Waters with a guard column (5 mm ˆ 2.1 mm). Eluent A was water/0.1% formic acid, and eluent B was acetonitrile/0.1% formic acid. For the analysis of the green reaction products, the following gradient program was used: 0 min, 2% B; 20 min, 17.7% B; 20.5 min, 100% B; 22.5 min, 100% B; 23 min, 2% B; 25 min, 2% B. The flow rate was 0.4 mL/min. MS parameters were as follows: capillary voltage, 1.5 kV; cone voltage, 25 V; extractor voltage, 2.0 V; RF voltage, 0.1 V; source temperature, 150 °C; desolvation temperature, 450 °C; cone gas (nitrogen), 50 L/h; desolvation gas (nitrogen), 900 L/h. Monitoring was performed at 450 nm, representing the approximate absorption maxima of the resulting adducts, and the UV-Vis spectra were recorded from 190–700 nm (peak width 0.2 min). Negative ion mass spectra of the column eluate were recorded in the range *m/z* 500–2000. For the product ion scans, argon was used as the collision gas (0.22 mL/min).

4   Conclusions

We subjected SEM to alkaline treatment and obtained evidence of the formation of several green trihydroxy benzacridine derivatives using UHPLC-DAD-MS/MS analysis. For the first time, the formation of the same TBA isomers in model systems containing CQA and individual amino acids at pH 9 was demonstrated, regardless of the amino acid employed. During oxidation by polyphenoloxidase at pH 7.0, only three of these TBA isomers were found. Contrary to the formation of TBA derivatives reported by Namiki et al. and other groups, neither peptide nor amino acid residues were attached to the resultant benzacridine core, except for model systems containing β-alanine or lysine. Whether any amino acid residues remain attached to the TBA cores seems to depend on which NH_2 group is involved in TBA formation. Taking together these results, the formation of TBA derivatives might be caused by the reaction between CQA quinones and free NH_2 groups. Our findings show that food protein interactions with phenolic compounds are still poorly understood. With attention on polyphenolic compounds as potential food additives, further research needs to be dedicated to the underlying interaction mechanism and implications of these findings in food and feed quality and safety.

References

Ali, M.; Homann, T.; Khalil, M.; Kruse, H.-P.; Rawel, H. Milk whey protein modification by coffee-specific phenolics: Effect on structural and functional properties. J. Agric. Food Chem. 2013, 61, 6911–6920.

Ali, M.; Homann, T.; Kreisel, J.; Khalil, M.; Puhlmann, R.; Kruse, H.-P.; Rawel, H. Characterization and modeling of the interactions between coffee storage proteins and phenolic Compounds. J. Agric. Food Chem. 2012, 60, 11601–11608.

Clifford, M.N.; Johnston, K.L.; Knight, S.; Kuhnert, N. Hierarchical scheme for LC-MSn identification of chlorogenic acids. J. Agric. Food Chem. 2003, 51, 2900–2911.

García Cárnovas, F.G.; Sánchez, J.V.; Iborra Pastor, J.L.; Lozano Teruel, J.A. The role of pH in the melanin biosynthesis pathway. J. Biol. Chem. 1982, 15, 8738–8744.

González-Pérez, S.; Merck, K.B.; Vereijken, J.M.; van Koningsfeld, G.A.; Gruppen, H.; Voragen, A.G.J. Isolation and characterization of undenatured chlorogenic acid free sunflower (Helianthus annuus) proteins. J. Agric. Food Chem. 2002, 50, 1713–1719.

Kammerer, D.R.; Saleh, Z.S.; Carle, R.; Stanley, R.A. Adsorptive recovery of phenolic compounds from apple juice. Eur. Food Res. Technol. 2007, 224, 605–613.

Kroll, J.; Rawel, H.M.; Rohn, S. Reactions of plant phenolics with food proteins and enzymes under special consideration of covalent bonds. Food Sci. Technol. Res. 2003, 9, 205–218.

Le Bourvellec, C.; Renard, C. Interactions between polyphenols and macromolecules: Quantification methods and mechanisms. Crit. Rev. Food Sci. Nutr. 2012, 52, 213–248.

Namiki, M.; Yabuta, G.; Koizumi, Y.; Yano, M. Development of free radical products during the greening reaction of caffeic acid esters (or chlorogenic acid) and a primary amino compound. Biosci. Biotechnol. Biochem. 2001, 65, 2131–2136.

Pedrosa, M.M.; Muzquiz, M.; Garcia-Vallejo, C.; Burbano, C.; Cuadrado, C.; Ayet, G.; Robredo, L.M. Determination of caffeic and chlorogenic acids and their derivatives in different sunflower seeds. J. Sci. Food Agric. 2000, 80, 459–464.

Pierpoint, W. O-quinones formed in plant extracts: Their reactions with amino acids and peptides. Biochem. J. 1969, 112, 609–616.

Prigent, S.V.E. Interactions of Phenolic Compounds with Globular Proteins and Their Effects on Food-Related Functional Properties. Ph.D. Thesis, Wageningen University, Wageningen, The Netherlands, 2005.

Prigent, S.V.E.; Voragen, A.G.J.; Li, F.; Visser, A.J.W.G.; van Koningsfeld, G.A.; Gruppen, H. Covalent interactions between amino acid side chains oxidation products of caffeoylquinic acid (chlorogenic acid). J. Sci. Food Agric. 2008, 88, 1748–1754.

Rawel, H.M.; Meidtner, K.; Kroll, J. Binding of selected phenolic compounds to proteins. J. Agric. Food Chem. 2005, 53, 4228–4235.

Schilling, S.; Sigolotto, C.-I.; Carle, R.; Schieber, A. Characterization of covalent addition products of chlorogenic acid quinone with amino acid derivatives in model systems and apple juice by high-performance liquid chromatography/electrospray ionization tandem mass spectrometry. Rapid Commun. Mass Spectrom. 2008, 22, 441–448.

Weisz, G.M.; Schneider, L.; Schweiggert, U.; Kammerer, D.R.; Carle, R. Sustainable sunflower processing– I. Development of a process for the adsorptive decolorization of sunflower [Helianthus annuus L.] protein extracts. Innov. Food Sci. Emerg. Technol. 2010, 11, 733–741.

Yabuta, G.; Koizumi, Y.; Namiki, K.; Hida, M.; Namiki, M. Structure of green pigment formed by the reaction of caffeic acid esters (or chlorogenic acid) with a primary amino compound. Biosci. Biotechnol. Biochem. 2001, 65, 2121–2130.

Chapter 4: Concluding remarks

This study aimed at understanding the underlying interaction mechanisms between phenolic compounds and food proteins. In this context, special focus was given to a possible use of reaction products between polyphenols and single amino acids. For the animal feed industry, a direct exploitation of the covalent reaction between chlorogenic acid (CQA) quinone and protein might be a promising approach for the production of high-protein feed stuff from solvent-extracted sunflower meal (SEM), the efficient utilization of dietary crude protein (CP) in ruminant nutrition is crucial for an optimal supply of nitrogen to ruminal microbes and amino acids to the animal. Due to excessive microbial CP degradation in the rumen, the efficiency of protein utilization in the small intestine is highly decreased. For this reason, ruminants benefit greatly from ruminally undegraded CP sources that escape degradation but can be hydrolyzed, leaving the released amino acids for post-ruminal absorption. Previous attempts aimed at the preparation of high-CP sources include the use of polymeric coatings, heat, addition of ionophores, treatment of proteins with chemicals such as formaldehyde or different acids, and the addition of secondary plant metabolites. Unfortunately, each of these approaches has its disadvantages. For example, some phytogenic additives may result in a decreased feed intake due to sensory effects, and the use of chemicals has raised safety concerns. Maillard reactions (due to heat treatment) can be poorly controlled and may lead to decreased intestinal digestibility and, finally, bioavailability of some amino acids [Antoniewicz et al., 1992; Wu and Papas, 1997; Yu et al., 2002; Patra and Saxena, 2009; Calsamiglia et al., 2010; chapter 2].

By alkaline treatment, the *in vitro* intestinal digestibility of the SEM can be increased by 10%, and the CP values can be enhanced up to 12%. Interestingly, the modified SEM displayed a somewhat green color, intensifying with increasing pH. It appears that the alkaline treatment resulted in the oxidation of CQA to its *o*-quinone and the subsequent formation of green reaction products, which correspond to fragmentation patterns of tryhydroxy benzacridine (TBA) derivatives [chapter 3; chapter 2].

The exact mechanism behind the pigment coloration remains unknown, although a charge transfer complex between unequally oxidized forms of the trihydroxy benzacridine seems likely: the resulting quin-hydrones would create a bathochrome effect, thus inducing the perception of green color. Another explanation for this effect might simply lie in the semi-quinone or quinone form of the respective heterocyclic compound. Also, the nitrogen molecule in the ring structure of the derivative might result in the formation of a merocyanine structure and, consequently, in a bathochrome effect. A bound alkyl group would enhance this effect at the nitrogen molecule by serving as a donor [Prigent et al., 2003; 2005].

Namiki et al. (2001) and Yabuta et al. (2001) reported that, upon oxidation, CQA quinone gives rise to the formation of a dimer, which subsequently reacts with the amino compound to finally yield a green benzacridine derivative. CQA dimers are known to be the main compounds which react with amino acids, as also observed for caffeic acid esters by Namiki and co-workers (2001). In comparison, polyphenol-protein reactions with CQA monomers occur rather infrequently, owing to the higher redox potential of the monomers and, subsequently, their diminished reactivity. It should be noted that several structures have been suggested for caffeic acid and CQA dimers, but only a few have been structurally elucidated [Namiki et al.,2001; Yabuta et al., 2001; Prigent et al., 2008; Schilling et al., 2008; chapter 3].

For the formation of TBA derivatives, Yabuta et al. (2001) concluded that an *ortho*-diphenol structure with a carbonyl side chain as present in CQA quinone is crucial. Also, TBA derivatives seem to emerge exclusively from primary amino groups. Depending on which amino acid is involved, the color of the reaction product varies between light and dark green and, in the case of tryptophan, red. Particularly the reaction between a CQA-dimer and lysine, which possesses both an α-amino group and an ε-amino group, results in covalent reaction products of an intense green color. In contrast, glutamine, arginine, asparagine, methionine, serine and threonine show only little reactivity towards CQA quinones. The red color of the TBA derivative emerging from tryptophan and CQA can be explained by the presence of the indole group of tryptophan, which contributes to the mesomeric system, thereby shifting the absorption spectrum to a lower range. Cysteine is likely to react with CQA monomers instead of dimers, due to its highly nucleophilic character. Consequently, it does not form any colored adducts. In reaction mixtures containing 5-CQA, incubation at even comparatively low temperatures leads to the formation of 3-CQA and 4-CQA, which can be identified via UHPLC-DAD-MS/MS and which match the characteristic fragmentation pattern as described by Clifford et al. (2003). In addition, a large number of peaks of low intensity can be detected that correspond to CQA dimers as described by Schilling et al. [Yabuta et al., 2001; Clifford, 2003; Prigent et al., 2005; Prigent et al., 2007; Prigent et al., 2008, Schilling et al., 2008; chapter 3].

In alkalized SEM extract six peaks are dominant, although several more distinct peaks can be detected. Except for cysteine and proline, the same applies for model systems, regardless of the amino acid employed. Corresponding reaction mixtures containing PPO at neutral conditions yield three peaks that correspond to three of the six peaks in the alkaline solutions. All peaks display an *m/z* ratio of 700, matching fragments of corresponding reaction adducts as described by Schilling et al. (2008). Although all peaks with *m/z* 700 are nearly identical in terms of fragmentation, they are detected at different retention times, suggesting that the corresponding compounds are structural isomers [Clifford, 2003; Schilling et al., 2008; chapter 3].

These findings contradict Namiki et al. (2001), who described a reaction mechanism for the formation of TBA derivatives with the corresponding amino acid side chain attached to the resulting benzacridine core. It should be mentioned that Namiki et al. (2001) proposed the reaction mechanism on the basis of n-butylamine, whereas Schilling et al. (2008) used N-BOC-lysine, thereby blocking the α-NH$_2$ group of the amino acid [Namiki et al., 2001; Schilling et al., 2008; chapter 3].

In this study, the α-NH$_2$ group of all tested amino acids was available for the reaction, as proposed by Namiki et al. (2001) for *n*-butylamine. To verify that the reaction of the α-NH$_2$ group is accountable for the loss of the amino acid side chain, β-alanine was used in additional CQA model systems. Neither alkaline nor enzymatic treatment resulted in peaks with *m/z* 700. Instead, in the enzymatically treated β-alanine solution, three peaks with *m/z* 772 were detected, accounting for TBAs with an alanine side chain attached. Accordingly, in the alkaline-treated β-alanine solution, six peaks with *m/z* 772 were detected, which corresponded to the three or six peaks, respectively, shown in the respective α-alanine systems [chapter 3].

These findings suggest that the presence, or lack, of an amino acid residue in the resulting TBA core depends on which NH$_2$ group is involved in the TBA formation. If β- or ε-NH$_2$ groups are incorporated in the TBA core, it can be assumed that the amino acid residue remains attached. This presumption is confirmed by recent research regarding milk and coffee proteins, in which ε-NH$_2$ groups of lysine were found to be incorporated in proteins that had been modified with phenolics. The formation of TBA derivatives might be caused by a reaction between CQA quinones and free NH$_2$ groups [Ali et al., 2012; 2013; chapter 3].

For the last decade, great progress has been made in the recovery of functional foods from natural sources, with an increasing focus on food processing by-products, such as fruit peels or – as adressed in this study- SEM. Being a co-product from sunflower oil production, SEM is rich both in polyphenols and protein. Sunflower proteins display various functional characteristics comparable to those of soybean and other legume proteins, like emulsifying or foaming properties, which makes isolated SEM protein an potentially interesting and comparativley low-priced source for food additives [Weisz et al., 2010; chapter 2]. As mentioned earlier, the highest protein yields are usually obtained by alkali extraction, which results in the formation of CQA quinones due to alkaline oxidation. Interactions between CQA quinones and protein (or, respectively, free amino acids) may lead to the formation of covalent reaction products that are responsible for discolorations. Proteins that have been covalently modified by oxidized CQA are blue-green at neutral and alkaline pH, which is usually not an appealing color for most food products. However, apart from chlorophylle, green pigments rarely emerge in natural food systems. As chlorophylle is known for its instability during heat treatment, green products from CQA protein interactions might be an interesting alternative in the food industry [Yabuta et al., 2001 Prigent, 2005; chapter 3].

In a recent study, Iacomino et al. (2017) explored the potential of benzacridine derivatives as food coloring. In addition to TBA derivatives originating from the reaction of CQA with amino acids, similar pigments were obtained by an analogous reaction of CQA with bovine serum albumin, and chicken egg white. The cytotoxicity of the pigments, either purified or as part of the original protein matrices, was assessed by performing the 3-(4,5-dimethylthiazol-2-yl)-2,5-diphenyltetrazolium bromide (MTT) reduction inhibition assay. The assay was run on human epithelial colorectal adenocarcinoma cells (CaCo-2) and on human liver cancer cells (Hep G2) at different times of incubation. Neither the purified pigments from amino acids nor the pigmented protein mixtures exerted significant toxicity against two human cell lines, at doses compatible with common use in food coloring. In additional experiments, Iacomino et al. (2017) tested the potential of the purified benzacridine pigments for various food applications. Stability and color properties were evaluated in a series of model systems, and using alginate hydrogels, cow's milk and soy milk. The thermal stability over a range of temperature of relevance to industrial transformations or home cooking was reported to be satisfactory, at variance from what was observed in the case of other natural green coloring such as chlorophylls. Finally, the exploitation of benzacridine pigments was suggested to be of use for the sensing of volatile amines associated with microbial spoilage of food [Iacomino et al., 2017].

Taken together, a directed exploitation of polyphenol-protein reactions is an auspicious approach to adress several food related issues, including natural product-based green food coloring. With regard to a possible use of polyphenols as functional food additives, this study shows that protein interactions with phenolic compounds are still not entirely understood. Although SEM or other natural products might be a promising source for adapted ruminant feed or even food additives, further research needs to be dedicated to the underlying interaction mechanism and implications of these findings in food and feed quality and safety.

References

Ali, M.; Homann, T.; Kreisel, J.; Khalil, M.; Puhlmann, R.; Kruse, H.-P.; Rawel, H. (2012). Characterization and modeling of the interactions between coffee storage proteins and phenolic Compounds. J Agric Food Chem 60, 11601–11608.

Ali, M.; Homann, T.; Khalil, M.; Kruse, H.-P.; Rawel, H. (2013). Milk whey protein modification by coffee-specific phenolics: Effect on structural and functional properties. J Agric Food Chem 2013, 61 6911–6920.

Antoniewicz, A.; van Vuuren, A.; van der Koelen, C.; Kosmala, I. (1992): Intestinal digestibility of rumen undegraded protein of formaldehyde-treated feedstuffs measured by mobile bag and in vitro technique. Anim Feed Sci Technol 39, 111–124.

Calsamiglia, S.; Ferret, A.; Reynolds, C.; Kristensen, N.; van Vuuren, A. (2010) Strategies for optimizing nitrogen use by ruminants. Animal 4, 1184–1196.

Clifford, M.N.; Johnston, K.L.; Knight, S.; Kuhnert, N. (2003): Hierarchical scheme for LC-MSn identification of chlorogenic acids. J Agric Food Chem 51, 2900–2911.

González-Pérez, S.; Merck, K. B.; Vereijken, J. M.; Van Koningsveld, G. A.; Gruppen, H.; Voragen, A.G. (2002). Isolation and characterization of undenatured chlorogenic acid free sunflower (Helianthus annuus) proteins. J Agric Food Chem 50, 1713–1719.

González-Pérez, S.; Vereijken, J. M. (2007). Sunflower proteins: Overview of their physicochemical, structural and functional properties. J Sci Food Agric 87, 2173–2191.

Iacomino, M.; Weber, F.; Gleichenhagen, M.; Pistorio, V.; Panzella, L.; Pizzo, E.; Schieber, A.; d'Ischia, M.; Napolitano, A. (2017). Stable benzacridine pigments by oxidative coupling of chlorogenic acid with amino acids and proteins: Toward natural product-based green food coloring. J Agric Food Chem 65, 31, 6519–6528.

Lomascolo, A.; Uzan-Boukhris, E.; Sigoillot, J.-C.; Fine, F. (2012). Rapeseed and sunflower meal: a review on biotechnology status and challenges. Appl Microbiol Biotechnol 95, 1105–1114.

Namiki, M.; Yabuta, G.; Koizumi, Y.; Yano, M. (2001): Development of free radical products during the greening reaction of caffeic acid esters (or chlorogenic acid) and a primary amino compound. Biosci Biotech Biochem 65, 2131–2136.

Patra, A.; Saxena, J. (2009). Dietary phytochemicals as rumen modifiers: a review of the effects on microbial populations. Anton Leeuw Int 96, 363–375.

Prigent, S.V. (2005). Interactions of phenolic compounds with globular proteins and their effects on food-related functional properties. Dissertation. Wageningen University, Netherlands.

Prigent, S.V.; Voragen, A.G.; van Koningsveld, G.; Gruppen, H.; Visser, A. (2007). Covalent interactions between proteins and oxidation products of caffeoylquinic acid (chlorogenic acid). J Sci Food Agric 87, 2502–2510.

Prigent, S.V.; Li, Feng; van Koningsveld, G.; Voragen, A.; Visser, A.; Gruppen, H. (2008). Covalent interactions between amino acid side chains and oxidation products of caffeoylquinic acid (chlorogenic acid). J Sci Food Agric 88,1748–1754.

Rawel, H.M.; Meidtner, K.; Kroll, J. (2005). Binding of selected phenolic compounds to proteins. J Agric Food Chem 53, 4228–4235.

Schilling, S.; Sigolotto, C.-I.; Carle, R.; Schieber, A. (2008). Characterization of covalent addition products of chlorogenic acid quinone with amino acid derivatives in model systems and apple juice by high-performance liquid chromatography/electro-spray ionization tandem mass spectrometry. Rapid Commun Mass Spectrom 22, 441–448.

Weisz, G. M.; Kammerer, D. R.; Carle, R. (2009). Identification and quantification of phenolic compounds from sunflower (*Helianthus annuus* L.) kernels and shells by HPLC-DAD/ESI-MSn. Food Chem 115, 758–765.

Weisz, G. M.; Schneider, L.; Schweiggert, U.; Kammerer, D. R.; Carle, R. (2010). Sustainable sunflower processing – I. Development of a process for the adsorptive decolorization of sunflower [*Helianthus annuus* L.] protein extracts. Inn Food Sci Emerg Technol 11, 733–741.

Wu, S.; Papas, A. (1997). Rumen-stable delivery systems. Adv Drug Delivery Rev 28, 323–334.

Yabuta, G.; Koizumi, Y.; Namiki, K.; Hida, M.; Namiki, M. (2001). Structure of green pigment formed by the reaction of caffeic acid esters (or chlorogenic acid) with a primary amino compound. Biosci Biotech Biochem 10, 2121–2130.

Yu, P.; Goelema, J.; Leury, B.; Tamminga, S.; Egan, A. (2002). Using the DVE/OEB model to determine optimal conditions of pressure toasting on horse beans (*Vicia faba*) for the dairy feed industry. Anim Feed Sci Technol 99, 141–176.

Chapter 5: List of abbreviations

SEM	Solvent-extracted sunflower meal
CQA	chlorogenic acid, caffeoylquinic acid
PPO	polyphenoloxidase
CP	crude protein
N	nitrogen
CNCPS	Cornell net carbohydrate and protein system
ADICP	acid-detergent insoluble crude protein
TP	true protein
NDICP	neutral-detergent insoluble crude protein
NDF	neutral detergent fibre
TCA	trichloroacetic acid
CCD	central composite design
ANOVA	analysis of variance
SBM	soybean meal
CSM	cottonseed meal
TBA	trihydroxy benzacridine
N-BOC-lysine	N-butoxycarbonyl-lysine
PLE	Pressurized liquid extraction

Chapter 6: Summary

Phenolic compounds and proteins are important components in foods and food ingredients. Interactions of phenolic compounds with isolated, proline-rich proteins have been extensively studied. However, polyphenol interactions with food protein are still poorly understood. With attention on polyphenolic compounds or reaction products with food proteins as potential additives in foods and animal feed, this study aimed at understanding the underlying interaction mechanism and resulting interaction products. Special attention was given to the potential use of polyphenol-protein complexes in animal nutrition. For this purpose, alkaline treated solvent-extracted sunflower meal was subjected to diifferent procedures, all of which aimed at estimating the intestinal digestibility. It could be shown that a targeted alkaline treatment of naturally occurring compounds in feedstuffs might be a promising approach to provide ruminant feed with improved intestinal digestibility values.

The subsequent examination of corresponding model systems aimed at understanding the underlying interreaction mechanism between chlorogenic acid (CQA) and single amino acids as well as the identification of resulting interaction products. UHPLC-DAD-MS/MS analysis revealed that all samples contained the same green trihydroxy benzacridine (TBA) derivatives. Contrary to previous findings, neither peptide nor amino acid residues were attached to the resultant benzacridine core. The results indicate that the formation of TBA derivatives is caused by the reaction between CQA quinones and free NH_2 groups. The potential application of this reaction in food chemistry has not yet been addressed. Further research is necessary to elucidate the structure of the addition products for a comprehensive evaluation of food and feed safety aspects.

Chapter 7: **Zusammenfassung**

Polyphenole und Proteine sind Nahrungsbestandteile, deren Wechselwirkungen untereinander seit längerem im Fokus von Forschungsarbeiten stehen. Trotz wachsender Kenntnis auf diesem Gebiet gilt das Wissen um die genaue Natur von Interaktionen zwischen Polyphenolen und Protein noch als lückenhaft. Zwar ist vieles bekannt über die Reaktionsmechanismen zwischen Polyphenolen und isoliert vorliegenden, Prolin-reichen Proteinen, jedoch weiß man bislang wenig über entsprechende Interaktionen mit Nahrungsprotein. Mit Hinblick auf den gezielten Einsatz von Polyphenolen oder entsprechenden Reaktionsprodukten in der Ernährung von Mensch und Tier war es Ziel dieser Arbeit, zugrundeliegende Reaktionsmechanismen besser zu verstehen. Besonderes Augenmerk lag auf dem gezielten Einsatz von Polyphenol-Protein-Komplexen in Futtermitteln für Wiederkäuer. Eine alkalische Behandlung von Sonnenblumenextraktionsschrot erhöhte hierbei die von Haus aus geringe Verdauungsrate von Wiederkäuer-Futtermitteln signifikant. In entsprechenden Modelllösungen konnten per UHPLC-DAD-MS/MS-Analyse als maßgebliche Verbindungen grüne Trihydroxy-Benzacridin-Derivate identifiziert werden. Im Gegensatz zu bisherigen Erkenntnissen tragen diese jedoch weder Peptid- noch Aminosäurereste. Zusätzlich legen die Resultate dieser Arbeit nahe, dass Trihydroxy-Benzacridin-Derivate durch die Reaktion von Chlorogensäure-Quinonen mit freien NH_2-Gruppen entstehen. Der Fokus weiterer Forschungsarbeiten sollte nicht zuletzt auf Sicherheitsaspekten solcher Verbindungen für Human- und Tierernährung liegen.